OPTIMUM INDUCTIVE METHODS

SYNTHESE LIBRARY

STUDIES IN EPISTEMOLOGY,

LOGIC, METHODOLOGY, AND PHILOSOPHY OF SCIENCE

VOLUME 232

ROBERTO FESTA
Fellow of the Department of Philosophy of Science,
University of Groningen, The Netherlands

OPTIMUM INDUCTIVE METHODS

A Study in Inductive Probability, Bayesian Statistics, and Verisimilitude

KLUWER ACADEMIC PUBLISHERS
DORDRECHT / BOSTON / LONDON

Library of Congress Cataloging-in-Publication Data

Festa, Roberto.
Optimum inductive methods : a study in inductive probability, Bayesian statistics, and verisimilitude / by Roberto Festa.
p. cm. -- (Synthese library ; v. 232)
Includes bibliographical references and indexes.

1. Bayesian statistical decision theory. 2. Probabilities. 3. Induction (Mathematics) 4. Truth. I. Title. II. Series.
QA279.5.F47 1993
519.5'42--dc20 93-11840

ISBN 978-90-481-4318-4

Published by Kluwer Academic Publishers,
P.O. Box 17, 3300 AA Dordrecht, The Netherlands.

Kluwer Academic Publishers incorporates
the publishing programmes of
D. Reidel, Martinus Nijhoff, Dr W. Junk and MTP Press.

Sold and distributed in the U.S.A. and Canada
by Kluwer Academic Publishers,
101 Philip Drive, Norwell, MA 02061, U.S.A.

In all other countries, sold and distributed
by Kluwer Academic Publishers Group,
P.O. Box 322, 3300 AH Dordrecht, The Netherlands.

Printed on acid-free paper

Printed in the Netherlands

To my parents

TABLE OF CONTENTS

ACKNOWLEDGMENTS

My views on scientific inference have been decisively influenced by the works of Karl Popper and Rudolf Carnap. In particular, my research in this area has been inspired by Popper's view - that verisimilitude, or truthlikeness, is the main cognitive goal of science - and by Carnap's theory of inductive probabilities (and, more generally, by his Bayesian approach to scientific methodology).

In spite of the well known disputes between these two great masters of contemporary epistemology, I believe that a 'reconciliation' between Sir Karl Popper and Rudolf Carnap is possible. This book, indeed, is inspired by the conviction that Popper's verisimilitude view can be fruitfully embedded within the Bayesian approach.

Unfortunately there are very few epistemologists interested in inductive probabilities *and* verisimilitude. In particular, the set of students who have given important contributions to both fields seems to include only two members: Professor I. Niiniluoto and Professor T.A.F. Kuipers. Thanks to scholarships offered by the Italian Ministry of Foreign Affairs, I had the opportunity to work under the supervision of both.

In 1981-82 I was at the Department of Philosophy at the University of Helsinki where I worked on verisimilitude under Professor Niiniluoto's supervision (see Festa, 1982, 1983). I have been deeply influenced by Professor Niiniluoto's views on scientific inference and, more generally, on the methods and tools of philosophy of science. Among other things, I am grateful to Professor Niiniluoto for giving a sharp exposition of my unpublished proposals on verisimilitude in monadic first-order languages (Festa, 1982) in his book *Truthlikeness* (1987, pp. 319-321).

Afterwards, in 1983-84, I was at the Department of Philosophy at the University of Groningen where I continued my research into verisimilitude under Professor Kuipers's supervision (see Festa, 1986, 1987a).

In 1986-87 I started working, under Professor Kuipers's supervision, on the subject of this book (see Festa, 1987b and Festa and Buttasi, 1987). My personal debt to Professor Kuipers for the research leading to the completion of this book is enormous. His continuous support and encouragement during many years have been of the utmost importance to me. Indeed he was much more than the supervisor of this research: without him this book would never have been written.

At the beginning of 1990 Professor Kuipers suggested inviting Professor W. Molenaar, from the Department of Statistics and Measurement Theory of the Faculty of Behavioral and Social Sciences (University of Groningen), to join him in the supervision of my work. In spite of the clumsy statistical terminology used in the first draft of this book, Professor Molenaar very kindly accepted the invitation. His help in improving the form and the content of the earlier drafts has been invaluable. If a Bayesian statistician can appreciate the proposals advanced here without rejecting them immediately the credit is largely due to him.

Although I have profited enormously from the time and care that both my supervisors devoted to reading and criticizing the earlier drafts, this does not mean that they will be satisfied with everything that is in the final version. Of course any mistakes in it are mine, not theirs.

I also wish to thank Professor W. K. B. Hofstee, Professor I. Niiniluoto and Professor W. Schaafsma for reading the manuscript and suggesting several improvements.

I am most indebted to Professor A. Pasquinelli, Professor G. Sandri and Professor M. C. Galavotti of the Department of Philosophy at the University of Bologna. From all of them I received stimulus and encouragement while studying for my degree in philosophy and later.

My interest in Bayesian statistics and its relationships with the theory of inductive probabilities was strongly stimulated by Professor D. Costantini through his works, numerous talks and his comments on my first paper on the subject of this book (Festa, 1984).

I also have an intellectual debt to several friends and colleagues who have - through publications, correspondence, and personal contacts - influenced my work. In particular, I would like to mention Dario Antiseri, Alberto Artosi, Roger Cooke, Paolo Garbolino, Pierdaniele Giaretta, Giulio Giorello, Risto Hilpinen, Isaac Levi, Marco Mondadori, David Pearce, Marcello Pera, Gianguido Piazza, Claudio Pizzi, Yao H. Tan, Raimo Tuomela, Bas van Fraassen, Henk Zandvoort.

I am grateful to Professor L. Nauta of the Department of Philosophy at the University of Groningen, for his help and kindness. I also owe my thanks to Ms Mariëtte Elzenga and Mr Nico de Jager for their assistance while I was working at the library of the Department of Philosophy at the University of Groningen.

For practical help in preparing the manuscript, I wish to thank especially Ms Cheryl Gwyther (for improving the English style) and Mr Massimo Bonfà (for printing the camera-ready copy and drawing the figures).

The time I spent in Holland would not have been so enjoyable without the friendship of Hans, Simone, Annet, Claar, and Jos Mooij, Inge de Wilde, Barend van Heusden, Harrie Jonkman, Ineke Siersema and Jos Griffioen. I am particularly grateful to Claar Mooij and Barend van Heusden for the enormous practical help they gave me while I was working on the completion of this book, not only during my presence in Holland but also, and especially, during my absence.

CHAPTER 1

INTRODUCTION

This book deals with a basic problem arising within the *Bayesian approach* to scientific methodology, namely the choice of prior probabilities.[1] The problem will be considered with special reference to some inference methods used within *Bayesian statistics (BS)* and the so-called *theory of inductive probabilities (TIP)*.[2] In this study an important role will be played by the assumption - defended by Sir Karl Popper and the supporters of the current *verisimilitude theory (VT)* - that the cognitive goal of science is the achievement of a high degree of truthlikeness or verisimilitude.

A more detailed outline of the issues and objectives of the book is given in Section 1.

In Section 2 the historical background of the Bayesian approach and the verisimilitude theory is briefly illustrated. In Section 3, the methods used in TIP and BS for making *multinomial inferences* are considered and some conceptual relationships between TIP and BS are pointed out. In Section 4 the main lines of a new approach to the problem of the choice of prior probabilities are illustrated. Lastly, in Section 5 the structure of the book is described and a first explanation of some technical terms is provided.

1.1 AN OUTLINE OF ISSUES AND OBJECTIVES

According to the Bayesian approach the hypotheses[3] taken into examination by a scientist within a given empirical inquiry are appraised in terms of their posterior probabilities relative to the available data, where such posterior probabilities are obtained as follows:

(1) first the scientist's initial opinions about the hypotheses are formally represented by certain prior probabilities following the rules of the calculus of probability;

(2) subsequently the posterior probabilities of such hypotheses relative to the result of a given experiment are derived from their prior probabilities by applying Bayes's theorem.

The above mentioned probabilities, used to represent the scientist's *degrees of belief* in the hypotheses, will be called *epistemic probabilities.*[4] Furthermore, the term "*inductive probabilities*" will be often used w.r.t. the epistemic probabilities considered in TIP.

A troublesome point concerning (1) is the choice of the prior probabilities used to represent the scientist's initial opinions. The two 'traditional' views on this point are the *subjective view* and the *aprioristic view.*

According to the subjective view the axioms of the probability calculus are the only 'principles of rationality' ruling the choice of prior probabilities. Any choice is admissible as long as a given scientist feels that it corresponds with his initial opinions. More precisely, any scientist should select the prior probabilities best fitting his initial opinions.

In conflict with the subjective view, the different types of 'objectivism' share the assumption that the choice of prior probabilities is ruled - in addition to the probability axioms - by further principles of rationality. In particular, according to the *aprioristic view* a unique admissible choice can be specified on the basis of appropriate principles of rationality justified by a priori considerations.

In this book the two traditional views are rejected and a new 'objective view' - called the *contextual view* - is proposed. According to this view the choice of prior probabilities is ruled by appropriate *context-dependent principles of rationality* which can be applied only in certain 'cognitive contexts'. The notion of cognitive context can be illustrated as follows.

An empirical inquiry does not consist simply in collecting experimental data to (dis)confirm a given hypothesis. Indeed in any empirical inquiry an important role is played by different elements such as: (1) the background knowledge shared by the scientists, i.e., the corpus of theories and other assumptions taken for granted, at least within that inquiry; (2) the cognitive goals pursued by scientists *qua* scientists.[5] These elements form the *cognitive context* of the inquiry, i.e., the specific 'cognitive situation' where the inquiry is actually made.[6]

The notions of *background knowledge* and *cognitive goals* will play an important role in the development of the contextual view. More specifically, concerning cognitive goals, the following *verisimilitude thesis (VER)* - which represents a basic assumption of VT - will be adopted:

(VER) The main cognitive goal of science is the achievement of a high degree of truthlikeness or verisimilitude.

1.2 THE PROBABILISTIC AND THE VERISIMILITUDE VIEWS: TWO FALLIBILISTIC METHODOLOGICAL TRADITIONS

The Bayesian view of scientific methodology and the verisimilitude theory belong to two different methodological traditions which originated in the seventeenth and the eighteenth centuries from the crisis of infallibilism.

From antiquity to the first half of the eighteenth century *infallibilism* was the prevailing methodological view in philosophy of science.[7] According to this view the cognitive goal of science is to discover exactly true theories about the world and to achieve complete *certainty* about the *truth* of such theories.[8]

The *truth* represents the *ontic element* of infallibilism (since the truth of a theory concerns its relationships with reality) whilst *certainty* represents the *epistemic element* (since certainty can be seen as one of the possible 'epistemic attitudes' towards a theory).[9]

From the middle of the eighteenth century it became increasingly clear that infallibilism was indefensible and that more modest 'methodological ideals' were needed (see Laudan, 1973, pp. 277-278). For instance, the epistemic element or the ontic element of infallibilism could be weakened: certainty could be replaced by *probability* and the truth by *verisimilitude*. Indeed, both possibilities were explored and, accordingly, two different *fallibilistic views* of scientific knowledge were proposed.

In the middle of the seventeenth century the ancient tradition of 'probabilism', dating back to Carneades in the second century B. C., was revived by a number of philosophers and mathematicians including Pascal, Huygens and Jacques Bernoulli. Contrary to the infallibilistic tenet that certainty is the only epistemic state compatible with genuine science,[10] the *probabilistic view* suggests that in most cases scientists can only attain non-extreme degrees of belief in the truth of the considered hypotheses, i.e., degrees of belief represented by probabilities different from 0 and 1.[11]

In the middle of the eighteenth century a number of scientists and philosophers, such as David Hartley, George LeSage and Joseph Priestley, proposed a different fallibilistic view - henceforth referred to as the *verisimilitude view* - based on "the doctrine that scientific methods are self-corrective, that science in its development is inexorably moving closer to the truth by a process of successive approximation" (Laudan, 1973, p. 276).[12]

The probabilistic view and the verisimilitude view evolved into two distinct methodological traditions often in competition with each other. I am inclined to believe that the rivalry between these two fallibilistic traditions is due more to historical reasons than to theoretical incompatibility and that a *'mature' fallibilistic methodology* can be developed by integrating the two traditions.[13] For instance, the contextual view of prior probabilities proposed in this book can be seen as a 'co-ordinated development' of TIP, BS and VT where the first two methodological theories belong to the tradition of probabilistic fallibilism, and the third belongs to the tradition of verisimilitude fallibilism.[14]

1.3 BAYESIAN STATISTICS AND THE THEORY OF INDUCTIVE PROBABILITIES

Empirical sciences make use of different kinds of inferences leading from a set of *premises* to another statement, usually called *conclusion.*[15] According to a traditional distinction inferences are divided into deductive and inductive.

While in deductive inferences the information conveyed by the conclusion is already included, more or less explicitly, in the premises, in *inductive inferences* the conclusion says something *more* - or something *new* - w.r.t. the premises. Hence, even though the truth of the premises is taken as guaranteed (e.g. when the premises describe some experimental evidence) one cannot be completely certain of the truth of the conclusion, since there is inevitably the risk that the 'additional information' contained in it is false.[16]

A particular kind of inductive inference is given by *multinomial inferences*, which can be described as follows.

Suppose that $k \geq 2$ mutually exclusive and jointly exhaustive categories $Q_1,...,Q_k$ are used to classify the outcomes of any trial of an experimental process *Ex*. Then, if there is a *physical probability*[17] $\mathbf{q}_i$ that the outcome of an arbitrary trial of *Ex* is Q_i (for $i = 1,...,k$) and this probability is constant across trials (neither influenced by trial number nor by the outcomes of previous trials), then *Ex* may be referred to as a Bernoulli process (if $k = 2$) or a multivariate Bernoulli process (if $k > 2$). In the case where the scientist's background knowledge includes the assumption that the investigated 'multicategorical' process *Ex* is a (multivariate) Bernoulli process, the scientist's cognitive context will be called *multinomial context.* Although typically a scientist does not know the true value of the parameter vector $\mathbf{q}=(\mathbf{q}_1,...,\mathbf{q}_k)$ governing *Ex*, he can formulate different hypotheses about $\mathbf{q}$ or the results of future trials of *Ex*. The inductive inferences concerning such hypotheses will be called *multinomial inferences.*

As seen above, according to the Bayesian view the scientist's degrees of belief in the hypotheses under examination can be represented by suitable epistemic probabilities. For instance, given the evidence *e* describing a sequence of trials of *Ex* and a hypothesis *h* concerning **q**, or future trials of *Ex*, a Bayesian multinomial inference is given by the determination of the posterior probability $p(h/e)$ of *h* relative to *e*.

In particular, in BS the starting point of multinomial inferences is to provide an adequate representation of the scientist's initial opinions about the value of **q**. For this purpose a so-called cumulative distribution function $F(q_1,...,q_{k-1})$, defined for all the values $(q_1,...,q_{k-1})$ of the parameter vector $(\mathbf{q}_1,...,\mathbf{q}_{k-1})$, is used. (The last component $\mathbf{q}_k$ follows from the requirement that the sum of all *k* probabilities equals 1: see Chapter 3.4.) From $F(q_1,...,q_{k-1})$ - which is frequently called *prior distribution* - the prior probability $p(e_n)$ of any sequence e_n of outcomes of *n* trials of *Ex* can be derived. Moreover, given the evidence *e* describing one or more earlier trials of *Ex*, the prior distribution $F(q_1,...,q_{k-1})$ can be updated by Bayes's theorem in order to obtain the posterior distribution $F(q_1,...,q_{k-1}/e)$ which allows one to derive the posterior probability $p(h/e)$ of any hypothesis *h* concerning **q** or future trials of *Ex*.

The *inductive methods* introduced in TIP can be seen as procedures to determine the epistemic probabilities $p(h/e)$ for couples (h,e) of statements relative to multivariate Bernoulli processes and other kinds of multicategorical processes.[18] Usually an inductive method *I* is introduced by defining the corresponding *prior predictive distribution* p_I which specifies the prior probability $p_I(e_n)$ for any sequence e_n of outcomes of *n* trials of the considered process *Ex*. Afterwards, the requested probabilities $p(h/e)$ can be derived from p_I.

An inductive method *I* is called *exchangeable* iff, for any *n* and any e_n, the prior probability $p_I(e_n)$ depends *only* on the number of outcomes in e_n belonging to each of the categories $Q_1,...,Q_k$ and not on the order in which they occur in e_n. Somewhat surprisingly, it follows from *de Finetti's representation theorem* that the prior predictive distribution p_I corresponding to an exchangeable inductive method *I* can be derived from a unique prior distribution $F(q_1,...,q_{k-1})$. This signifies that the *exchangeable* inductive methods considered in TIP are 'equivalent' to the prior distributions used in BS for making multinomial inferences. In particular, it can be proved by using de Finetti's theorem that the well known exchangeable inductive methods proposed by Carnap and Stegmüller (1959) - henceforth referred to as *GC-systems*[19] - are equivalent to the so-called prior *Dirichlet distributions* commonly used in BS.

1.4 OPTIMUM PRIOR PROBABILITIES: THE CONTEXTUAL APPROACH

In this book the choice of prior probabilities will be considered with particular reference to multinomial contexts and, more specifically, to the problem of the choice of a GC-system for such contexts.

A GC-system can be referred to as (γ,λ) since it is fully characterized by a *prior vector* $\gamma=(\gamma_1,...,\gamma_k)$ and a *parameter* λ $(0 < \lambda < \infty)$ where: (1) each component γ_i $(i = 1,...,k)$ of γ denotes the prior probability of the hypothesis that the outcome of the first trial of the investigated process *Ex* is Q_i; (2) λ can be seen as an index of the reluctance of a subject using (γ,λ) to change his prior probabilities in response to the experimental evidence.[20]

The problem of the choice of a GC-system for a given multinomial context will be called the *epistemic problem of optimality (EPO)* and the selected GC-system will be referred to as *the optimum GC-system* $(\gamma°,\lambda°)$. The term "optimum" is suggested by the idea - underlying the approach to EPO developed in Chapter 8 - that a possible reason for selecting a given inductive method, from a certain class of inductive methods, is that there are grounds to believe that it is 'the optimum tool' for achieving a high degree of verisimilitude.

EPO can be split in two subproblems: (i) the selection of the optimum prior vector $\gamma°$, (ii) the selection of the optimum GC-system $(\gamma°,\lambda°)$ from the set of all the GC-systems $(\gamma°,\lambda)$ with prior vector $\gamma°$. The suggested solution to EPO - which can be seen as a particular application of the contextual view of prior probabilities (see Section 1) - applies to multinomial contexts where:

(1) The *background knowledge* shared by the scientists engaged in the inquiry includes an 'informal estimate' $Est[G(\mathbf{q})]$ of the (unknown) *Gini diversity* $G(\mathbf{q}) = 1 - \Sigma \mathbf{q}_i^2$ of *Ex*;[21]

(2) the *cognitive goal* pursued by the scientists is to minimize the distance between their (γ,λ)-based estimate of $\mathbf{q}$ and the truth (the true value of $\mathbf{q}$).

Given the above mentioned equivalence between GC-systems and prior Dirichlet distributions, solving EPO, i.e., choosing a GC-system for a given multinomial context, amounts to choosing the corresponding prior Dirichlet distribution. Hence, the suggested solution to EPO can be seen as a contribution to an issue which has been extensively investigated in BS.

However, it should be pointed out that the aims of this book *do not* include the foundation of BS, let alone the foundation of statistics. Indeed multinomial inferences are only one of many kinds of statistical inferences and the results obtained for multinomial inferences are not directly generalizable to the other cases. My purposes w.r.t. statistical practice are neither descriptive nor prescriptive. Far from saying what statisticians do or should do, I just suggest what Bayesian statisticians could do in a very idealized kind of cognitive context, i.e., multinomial contexts.[22]

1.5 THE LAYOUT OF THE BOOK

The prospective audience of this book includes readers interested in: (1) the theory of inductive probabilities (TIP); (2) the problem of the choice of prior probabilities in Bayesian statistics (BS); and (3) the verisimilitude theory (VT). More generally, it includes anyone interested in statistics and the philosophy of science. Given that many of these readers may be relatively unfamiliar with any of the three above mentioned subjects, the first part of the book (Chapters 2-4) provides introductory material on TIP, BS and VT.

In Chapter 2 some basic features of TIP are considered. In particular, some basic requirements satisfied by different types of inductive methods are illustrated and a concise classification of the 'intended applications' of such methods is given.

In Chapter 3 some basic features of BS are described and some methods used in BS for making multinomial inferences are briefly illustrated. A reader familiar with BS can skip the first part of the Chapter (Sections 1-3) and examine only the second part (Sections 4-5) where new 'predictive interpretations' of some well-known typical values of the probability distributions on the parameter vector of a multivariate Bernoulli process are suggested and some relatively unknown typical values - such as the expected Gini diversity - are considered.

In Chapter 4 a well known Bayesian point estimator, i.e., the mean estimator, is considered and a 'verisimilitude justification' of this estimator is suggested. The principal objective of this chapter is to prove that epistemic probabilities satisfy the so-called *principle of immodesty* which plays a very important role in the contextual approach to EPO. This chapter also provides a brief outline of the basic issues of the verisimilitude theory.

The main purpose of the second part of the book (Chapters 5-6) is to illustrate the basic features of GC-systems and Dirichlet distributions and the relationships between the two.

In Chapter 5 some implications of de Finetti's representation theorem are described. In particular, the one-to-one correspondence between the set of exchangeable inductive methods and the set of prior distributions on the parameter vector **q** of a multivariate Bernoulli process is considered. Moreover, some problems regarding the convergence of opinion in scientific research are examined. For instance, it is shown that exchangeable inductive methods normally allow scientists to learn from experience, not only in the 'epistemic sense' that scientists using different inductive methods are led to a considerable convergence of opinion, but also in the 'realist sense' that the predictive probabilities derived from such methods tend, in the long run, to approach the truth.

In Chapter 6 the basic features of GC-systems and Dirichlet distributions are considered and the relationships between the two are elucidated. More specifically, it is shown that there is a one-to-one correspondence between the class of GC-systems and the class of the prior Dirichlet distributions on the parameter vector **q** of a multivariate Bernoulli process. Moreover, some axiomatizations of GC-systems and Dirichlet distributions are considered.

The contextual view of prior probabilities and, in particular, a contextual approach to the epistemic problem of optimality (EPO) are illustrated in the third part of the book (Chapters 7-11).

In Chapter 7 the two 'traditional' views about the problem of the choice of prior probabilities, i.e., the subjective view and the aprioristic view, are considered and a relatively new contextual view is presented. In particular, a context-dependent justification for using a prior Dirichlet distribution (or, equivalently, a GC-system) for a given inquiry is suggested.

Chapter 8 contains the main technical results of the book. A contextual approach to EPO - i.e., the problem of determining the optimum Dirichlet distribution (or, equivalently, the optimum GC-system) for investigating a given multivariate Bernoulli process - is proposed. In particular, a *verisimilitude solution*, or *V-solution*, to EPO based on the verisimilitude thesis (VER) is given. The V-solution to EPO is obtained using the following strategy: (a) having defined the so-called *logical problem of optimality*, or *LPO*, for the set DIR(γ) of all Dirichlet distributions with a given prior vector γ, a V-solution to LPO is provided (Section 4); (b) using (VER) and the V-solution to LPO, a V-solution to EPO is elaborated (Section 5). Among other things it is shown (Section 7) that the V-solution to LPO represents a generalization of some results concerning the 'logically optimum inductive methods' obtained by Carnap (1952).

In Chapter 9 some features of the contextual approach to EPO are elucidated by comparing this approach with others.

In Chapter 10 Gini diversity, which plays a crucial role in the contextual approach to EPO, is considered in some detail.

Finally, in Chapter 11, the relationships between the contextual view and some recent methodological programmes of research are considered and a number of possible developments of the contextual approach to EPO are suggested.

How to use this book. The book uses the standard notation from elementary mathematics and assumes no previous knowledge on the part of the reader. Sections, formulae, figures and notes are numbered according to their order of appearance in each chapter. In the same chapter they are referred to simply by their numbers. In a chapter different from that where a formula is originally given, the formula number is preceded by its chapter number: e.g., formula 23 stated first in Chapter 5 is referred to as 'formula (5.23)'.

PART I

INDUCTIVE PROBABILITIES, BAYESIAN STATISTICS, AND VERISIMILITUDE

CHAPTER 2

THE THEORY OF INDUCTIVE PROBABILITIES: BASIC FEATURES AND APPLICATIONS

In this chapter some basic features and applications of the *theory of inductive probabilities (TIP)* are described.

In Section 1 the notion of *inductive method* is introduced and a number of fundamental requirements, satisfied by different types of *inductive methods*, are illustrated. In Section 2 a concise classification of the 'intended applications' of inductive methods, i.e., *multicategorical inferences*, is given. Lastly, in Section 3 the so-called *multinomial contexts*, where multicategorical inferences are frequently made, are defined.

2.1 INDUCTIVE METHODS

A set **Q** including *k* mutually exclusive and jointly exhaustive *categories*, or *properties*, $Q_1,...,Q_k$ can be used to describe (the composition of) a population *U* or samples of individuals randomly drawn from *U*. More generally **Q** can be used to describe an experimental process *Ex* or sequences of trials of *Ex*. The statements recording these descriptions will be referred to as *multicategorical statements*.

Multicategorial statements may refer to different kinds of multicategorial processes (see Chapter 11, note 4). The objective of TIP is to define appropriate *inductive methods* for such processes, i.e., appropriate procedures to determine the inductive probabilities $p(h/e)$ for all couples (h,e) of multicategorical statements or, at least, for a wide set of such couples.

Given a sequence e_n of outcomes of *n* trials of a multicategorical process *Ex*, the number of outcomes in e_n belonging to the category Q_i, or *Q_i-outcomes*, will be referred to as $n_i(e_n)$. Two sequences e_n and e'_n are called *structurally identical* ($e_n =_S e'_n$) iff, for any Q_i, $n_i(e_n) = n_i(e'_n)$.

The symbol "e_n" will also be used to denote a *complete description* of the sequence e_n, i.e., a statement specifying to which category each outcome in e_n belongs.

The hypotheses: "the outcome of the next trial of *Ex* will be Q_i", "the outcomes of the next two trials will be Q_i", and "the outcomes of the next two trials will be Q_i and Q_j" will be referred to as Q_i, Q_iQ_i, and Q_iQ_j, respectively, and their inductive probabilities relative to the evidence e describing a sequence of trials of *Ex* as $p(Q_i/e)$, $p(Q_iQ_i/e)$, and $p(Q_iQ_j/e)$. Lastly, the evidence obtained by adding the observation of a Q_i-outcome to e will be referred to as eQ_i.

A *prior predictive distribution* on *Ex* defines the prior probability $p(e_n)$ for any sequence e_n of outcomes of n trials of *Ex*. In particular, given an inductive method I, the prior predictive distribution specified by I will be referred to as p_I. An inductive method I is virtually identical to its prior predictive distribution p_I since, in general, the inductive probabilities $p(h/e)$ can be derived from p_I (cf. Carnap, 1952, p. 18). Hence, the expressions "prior predictive distribution on *Ex*" and "inductive method (for *Ex*)" will be used interchangeably.

In TIP the inductive probabilities $p(h/e)$ determined by any inductive method I are required to satisfy the axioms of probability calculus (cf. Carnap, 1971, p. 40):

(A1) *Lower Bound.* $p(h/e) \geq 0$
(A2) *Self-confirmation.* $p(e/e) = 1$
(A3) *Complement.* $p(h/e) + p(\neg h/e) = 1$
(A4) *Multiplication.* $p(h \,\&\, h'/e) = p(h/e)p(h'/e \,\&\, h)$

In (A1)-(A4) it is assumed that the second argument of function p is logically possible. Notice that (A1)-(A4) are stated in terms of the *conditional* probability function $p(h/e)$. The *absolute* probability function $p(h)$ is given by the definition "$p(h) = p(h/z)$" where z denotes the tautological proposition.

It follows from the multiplication rule (A4) that $p(e_nQ_i) = p(e_n)p(Q_i/e_n)$. The repeated application of this equality shows that the prior probability $p(e_n)$ of any sequence e_n can be obtained from the so-called *special values* $p(Q_i/e_n)$ including the initial ones $p(Q_i)$. For instance, the prior probability of $e_4 = Q_1Q_3Q_3Q_2$ is given by $p(e_4) = p(Q_1)p(Q_3/Q_1)p(Q_3/Q_1Q_3)p(Q_2/Q_1Q_3Q_3)$. This means that the special values allow one to determine a predictive distribution on *Ex* or, equivalently, an inductive method for *Ex* (cf. Carnap, 1952, pp. 16-18).

In addition to the axioms of probability calculus, other assumptions - or, to use the terminology of analytical philosophy, other *requirements of adequacy* - have been suggested in TIP to characterize different types of inductive methods. Some of these will be given below.

The *requirement of exchangeability (Exc)* states that identical inductive probabilities should be attributed to two structurally identically sequences of outcomes:

(Exc) If $e_n =_S e'_n$ then $p(e'_n) = p(e_n)$.[1]

An inductive method satisfying (Exc) is called *exchangeable*.

Another requirement of adequacy considered in TIP is the *requirement of regularity (Reg)* which demands that any sequence e_n should receive a positive probability:

(Reg) $p(e_n) > 0$

Finally, let us mention three requirements which might be considered *inductive* requirements in the strictest sense, since they concern the *effect of experimental evidence* on the inductive probabilities attributed to certain hypotheses.

The first of these 'strictly inductive' requirements is the *requirement of positive relevance (PR)* which demands that, whenever the empirical evidence e is enriched by the observation of a further Q_i-outcome, the inductive probability attributed to the hypothesis that the next trial be Q_i should *increase*:

(PR) $p(Q_i/eQ_i) > p(Q_i/e)$

By replacing ">" with "≥" in (PR) the weaker *requirement of non-negative relevance (NNR)* is obtained:

(NNR) $p(Q_i/eQ_i) \geq p(Q_i/e_n)$

The third strictly inductive requirement is the so-called *axiom of convergence*, or *Reichenbach axiom (Reich)* (cf. Gaifman, 1971, p. 244):[2]

(Reich) For every $\varepsilon > 0$ there exists N such that, for every e_n, if $n > N$ then $|p(Q_i/e_n) - n_i/n| < \varepsilon$, for $i = 1,...,k$.

Put simply, (Reich) requires that "as the total size of the sample becomes larger, the [probability $p(Q_i/e_n)$] that, given the sample, the next trial will belong to the i-th attribute, should approach the relative frequency [n_i/n] of this attribute within the sample" (Gaifman, 1971, pp. 243-244).

The above-mentioned requirements of adequacy are *predictive* requirements in the sense that they concern the inductive probabilities attributed to the 'predictions' of certain future events. In particular, the above requirements are stated, or can be restated, in terms of the special values $p(Q_i/e_n)$ which represent the inductive probabilities of certain 'singular predictions'. For instance, (Reg) is logically equivalent to the following requirement (cf. Kuipers, 1980, p. 185):[3]

(Reg)* $p(Q_i/e_n) > 0$

The specification of a certain class of inductive methods on the basis of a corresponding set of predictive requirements relative to special values has noticeable intuitive, philosophical and technical implications.

From an *intuitive* point of view, this strategy is convenient as it would appear much easier to explore our presystematic intuitions concerning the special values $p(Q_i/e_n)$ than, for instance, our intuitions regarding the prior probability $p(e_n)$ of certain finite sequences of experimental results.

From a *philosophical* point of view the special emphasis on special values springs from the idea, going back at least to John Stuart Mill, that inductive inferences are essentially inferences from particulars to particulars.[4]

Lastly, from a *technical* point of view the importance of special values is mainly related to the fact that they allow one to determine the predictive distribution p_I which corresponds to a given inductive method I (see above).

2.2 MULTICATEGORICAL INFERENCES

Let the evidence e in $p(h/e)$ describe a sample drawn from a given population U. Then, depending on the nature of h, one can distinguish different types of multicategorical inferences.[5]

For instance, one may distinguish between predictive and global inferences. In a *predictive inference*, the hypothesis h describes a sample to be taken from U. In particular, in a *singular predictive inference* the sample described by h consists of only one trial. In a *global inference*, h describes certain 'global features' of U.[6] A special type of global inference is the *universal inference*, where h is a universal hypothesis, for example, "All the individuals in U belong to one of the categories $Q_{i1},...,Q_{ig}$", where $Q_{i1},...,Q_{ig}$ is a subset of $\mathbf{Q}$.

Another important type of global inference is the *statistical inference* where h is a 'statistical hypothesis' concerning the value of N_i/N and N_i is the number of individuals in U belonging to the category Q_i.

For instance, one might try to calculate the inductive probabilities $p(h/e)$ and $p(h'/e)$ of the statistical hypotheses h = "$N_1/N = 0.3$" and h' = "$0.3 \leq N_1/N \leq 0.7$" relative to evidence e. In principle, such probabilities can be derived from the special values of a given inductive method by repeated application of the multiplication rule and other axioms of the probability calculus. In practice, however, when U includes a very large number N of individuals, such calculations become enormously laborious and the difficulties appear to increase when U is infinite (since, in this case, the field of the possible values of N_1/N does not include only (0, $1/N$, $2/N$,...,1) but is identical to the interval [0,1] including a non-denumerable infinity of values).[7]

2.3 MULTINOMIAL CONTEXTS

Multicategorical inferences are frequently made in those cognitive contexts which might be called ***multinomial (cognitive) contexts***.

Before introducing multinomial contexts, the so-called *multivariate Bernoulli processes* should be considered:

(1) Let *Ex* be an experimental process with $k > 2$ possible outcomes $Q_1,...,Q_k$ for any trial. Then *Ex* is a ***multivariate Bernoulli process*** iff:

(i) the objective probability of obtaining Q_i (with $i = 1,...,k$) in any trial is *independent* of the outcomes obtained in previous trials, and

(ii) it is *stable* over the sequence of trials.

It follows from (1) that the objective probability of Q_i in a multivariate Bernoulli process *Ex* is given by a fixed value q_i and, consequently, the objective probability $P_q(e_n)$ of a given sequence e_n of outcomes of n trials is given by:

$$(2) \qquad P_q(e_n) = q_1^{n_1} q_2^{n_2} \cdots q_k^{n_k} = \prod_{i=1}^{k} q_i^{n_i}$$

where n_i denotes the number of Q_i-outcomes in e_n. Accordingly, a multivariate Bernoulli process is characterized by an 'exchangeable distribution of outcomes', in the sense that the objective probability distribution $P_q(e_n)$ *does not* depend on the particular order in which $Q_1,...,Q_k$ occur in e_n, but on their frequencies $n_1,...,n_k$ only. This implies that identical objective probabilities are attributed to two sequences e_n and e_n' which are structurally identical:

(3) If $e_n =_S e_n'$ then $P_q(e_n') = P_q(e_n)$.

In particular, it follows from (2) that:

(4) (a) $P_q(Q_i) = q_i$
(b) $P_q(Q_iQ_i) = q_i^2$

where $P_q(Q_i)$ is the objective probability that the outcome Q_i in the next trial of *Ex* and $P_q(Q_iQ_i)$ is the objective probability of a sequence Q_iQ_i of two Q_i-outcomes.

The set of sequences which are structurally identical to a given sequence e_n will be referred to as $\ddot{e}_n$. If $n_1,...,n_k$ is the k-tuple of the frequencies of $Q_1,...,Q_k$ in e_n, then $\ddot{e}_n$ includes $n!/(n_1!n_2! \cdots n_k!)$ sequences. Hence, it follows from (2) and from (3) that the objective probability $P(\ddot{e}_n)$ of $\ddot{e}_n$ is given by:

(5) $$P(\ddot{e}_n) = n!/(n_1!n_2! \cdots n_k!) \prod_{i=1}^{k} q_i^{n_i}$$

The probability distribution in (5) is usually called the *multinomial distribution*.

Multivariate Bernoulli processes can be seen as an extension of the more familiar *Bernoulli processes* which are obtained from (1) by replacing "$k > 2$" by "$k = 2$".

Although complete certainty that a given real-world experimental process *Ex* is a (multivariate) Bernoulli process can never be guaranteed, in many cases it can be assumed, on the basis of the available background knowledge, that *Ex* will behave, at least approximately, as a (multivariate) Bernoulli process.

A typical application of the 'Bernoulli model' is given by the successive tosses of an apparently regular coin, where it is typically assumed that the objective probability q_1 of the appearance of the head is 1/2 in each toss. The Bernoulli model is also applied when the coin is visibly irregular. Although in this case the value of the objective probability $\mathbf{q}_1$ governing the process cannot be specified it can be estimated on the basis of a sequence of trials (see Chapter 3.5).

Consider an urn of constant composition containing N balls of k different colors $Q_1,...,Q_k$ where N_i is the number of balls of color Q_i in the urn. Then random sampling with replacement from the urn can be considered a multivariate Bernoulli process with objective probabilities $q_1,...,q_k$ where $q_i = N_i/N$. Where the composition of the urn is unknown, it can only be assumed that random sampling with replacement from the urn is a multivariate Bernoulli process, although in this case the value of the objective probabilities $\mathbf{q}_1, ..., \mathbf{q}_k$ which govern the process cannot be specified. The objective probabilities $\mathbf{q}_1,...,\mathbf{q}_k$ of a multivariate Bernoulli process *Ex* 'related' to a given population U may be directly referred to as the objective probabilities of $Q_1,...,Q_k$ in U.

The practical importance of the above 'urn model' cannot be over-estimated since (i) typically the composition of a multicategorical population is assumed to be (approximately) constant and (ii) the experimental process *Ex* consisting in the observation of individuals taken from the population is (approximately equal to) a random sampling with replacement.[8] Multinomial contexts can be defined as follows:

(6) Given an experimental process *Ex* with $k > 2$ possible outcomes $Q_1,...,Q_k$ for any trial, the cognitive context **C** is a *multinomial context* iff the background knowledge *BK* available in **C** includes the assumption that *Ex* is a multivariate Bernoulli process.[9]

"Multinomial context" will be used herein in a wide sense to include also those cases where *Ex* is a Bernoulli process ($k = 2$).

Although in a multinomial context the true value of the parameter vector $\mathbf{q}=(\mathbf{q}_1,...,\mathbf{q}_k)$ governing *Ex* is typically unknown, a researcher can, on the basis of the observations of a sequence of trials of *Ex*, make certain inductive inferences - which can be called *multinomial inferences* - concerning $\mathbf{q}$ or the results of future trials of *Ex*.

CHAPTER 3

BAYESIAN STATISTICS AND MULTINOMIAL INFERENCES: BASIC FEATURES

In this chapter some basic features of *Bayesian statistics (BS)* are described and some methods used in BS for making multinomial inferences are illustrated.

Section 1 includes a definition of BS and a taxonomy of Bayesian statistical inferences. In Section 2 some introductory notions of probability theory concerning univariate and multivariate probability distributions are given. In Section 3 the application of Bayes's theorem to global and predictive statistical inferences is illustrated.

Lastly, the probability distributions on the parameter vector $\mathbf{q}$ of a multivariate Bernoulli process *Ex* are considered (Section 4) and some methods used in BS for making multinomial inferences are examined (Section 5).

A reader familiar with BS can skip the first part of this chapter (Sections 1-3) and read only the second part (Sections 4-5) where new 'predictive interpretations' of some well-known typical values of the probability distributions on $\mathbf{q}$ are suggested and some relatively unknown typical values - such as the *expected Gini diversity* - are considered.

3.1 WHAT IS BAYESIAN STATISTICS?

A *statistical inference* is a particular type of inductive inference (see Chapter 1.3) where the empirical evidence concerns a sample taken from a given population and the inductive conclusion concerns the whole *population* or certain *samples* to be selected from it. More generally, a statistical inference is an inductive inference which concerns certain parameters of a given experimental *process* or *sequences* of trials of this process.

A *theory of statistical inference* provides a systematic corpus of procedures which should govern the different types of statistical inferences.

Although Good (1983b, p. 191), perhaps somewhat amusingly, says that there are at least 93312 varieties of Bayesians, all Bayesian statisticians presumably share a nucleus of views which represents, as it were, the *differentia specifica* distinguishing *Bayesian statistics (BS)* from the so-called orthodox statistics. The distinctive features of BS are the following:

(i) While orthodox statistics uses only empirical information concerning the sample, BS also takes account of the *prior information* researchers possess *before sampling*.
(ii) Prior information is formally specified by a distribution of *prior probabilities* on the possible values of certain *parameters* which characterize the population or process under investigation, where such probabilities are *epistemic*.
(iii) Using Bayes's theorem prior probabilities are updated on the basis of the empirical sample so as to obtain a distribution of *posterior probabilities* on the parameters. Posterior probabilities provide a formal representation of the researcher's beliefs *after sampling*.
(iv) The conclusions of a statistical inference are (derived from) the posterior probabilities.

Given a set of t mutually exclusive and jointly exhaustive hypotheses $h_1,...,h_t$ concerning the composition of a given population, the posterior probabilities $p(h_j/e)$ are derived from the prior probabilities $p(h_j)$ and the so-called *likelihoods* $p(e/h_j)$ by the following frequently used formulation of Bayes's theorem:

$$(1) \qquad p(h_j/e) = \frac{p(e/h_j)p(h_j)}{p(e)} = \frac{p(e/h_j)p(h_j)}{\sum_{i=1}^{t} p(e/h_i)p(h_i)}$$

The Bayesian strategy for combining prior information and new empirical information, so as to obtain the posterior probabilities, is represented in Figure 1 (cf. Zellner, 1971, p. 10).

Bayesian inferences may be classified using two different criteria.

According to the first criterium - relating to the 'target' of Bayesian inferences - a distinction should be made between *global* and *predictive inferences*:

(1a) *Global inferences* concern certain *global* features or parameters of the population or process under investigation.

(1b) *Predictive inferences* concern certain features of samples (sequences of trials) still to be drawn from the population (process) under investigation.

According to the second criterium - relating to the 'nature' of the conclusions of a statistical inference - a distinction should be made between *probabilistic* and *acceptational inferences*:

(2a) In *probabilistic inferences* the conclusions are the posterior probabilities of the different hypotheses considered.

(2b) In *acceptational inferences* the conclusions consist of the *(inductive) acceptance* of a hypothesis, while posterior probabilities play an important role in accepting that hypothesis.[1]

As far as the *rules of inductive acceptance* are concerned, one may distinguish between *pure* acceptational inferences, where the acceptance depends *exclusively* on the posterior probabilities of the hypotheses considered, and *utilistic* acceptational inferences, where the *utility* (*loss*) involved in the acceptance of each of the hypotheses in each of the possible

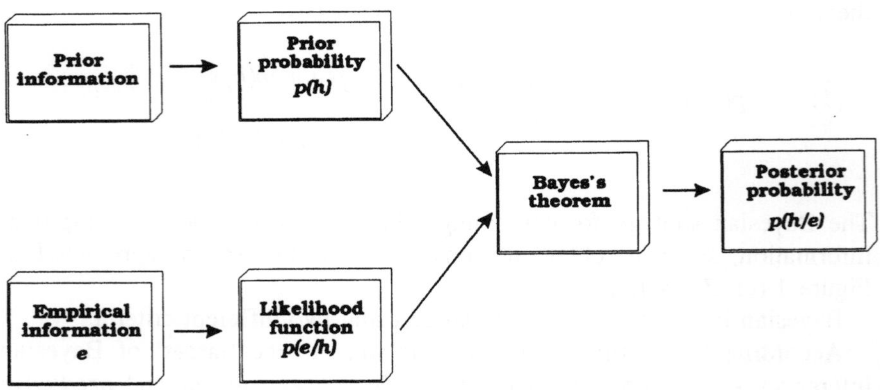

Figure 1

The Bayesian strategy for combining prior information and new empirical information

states of nature is also taken into account. In utilistic acceptational inferences various kinds of utilities may be employed. For instance, the so-called *decision-theoretic statistics* (cf. Ferguson, 1970) usually refers - more or less explicitly - to *pragmatic* utilities such as economical gains or losses. On the other hand, the so-called *theory of cognitive decisions* (cf. Levi, 1967, 1980 and Niiniluoto, 1987, Ch. 12) employs several *cognitive* utilities, such as truth, information content and verisimilitude.

On the basis of the methodological importance attributed to the various kinds of Bayesian inferences, it is possible to identify different types of Bayesian approaches.

Firstly consider the distinction between the *probabilistic approach* and the *acceptational approach*. According to the probabilistic approach the *only* information scientists really need are the posterior probabilities of hypotheses, while the acceptational approach emphasizes the methodological indispensability of acceptational inferences.[2]

Secondly consider the distinction between the *globalistic approach* and the *predictivistic approach*. According to the globalistic approach statistical inferences concern mainly the 'global features' of the population, while the predictivistic approach emphasizes the practical and scientific importance of predictive problems.[3] More specifically, the predictivistic approach stresses that the most important objective of statistical inferences consists in calculating the *posterior predictive distributions* $p(y/z)$, where y and z denote possible values of two given experiments **y** and **z**.[4]

3.2 PROBABILITY DISTRIBUTIONS

This section introduces the main features of the probability distributions of *unknown parameters*. Note, however, that all the included notions and formulae also apply to the probability distributions of random variables where a *random variable* is "a numerical quantity whose value is determined by an experiment of chance" (Lindgren, 1976, p. 50).

Kinds of probability distributions

The statistical inferences concerning the value of an unknown real-valued quantity - or *parameter* - θ characterizing a given population or process are called *univariate inferences*. For instance, θ may be (i) the proportion of defective pieces in a batch of industrially manufactured articles, (ii) the chance of an irregularly shaped coin turning up heads, (iii) the mean lifetime of a particular component of a mechanical instrument.

Bayesian univariate inferences are made on the basis of an appropriate probabilistic representation of the researcher's uncertainty regarding $\boldsymbol{\theta}$, i.e., of a *cumulative distribution function*, or *cdf*, $F(\theta)$. Given a value a of $\boldsymbol{\theta}$, the epistemic probability $p(\boldsymbol{\theta} \leq a)$ that $\boldsymbol{\theta}$ is lower than or equal to a is given by $F(a)$:

(2) $$F(a) = p(\boldsymbol{\theta} \leq a)$$

In the place of "cumulative distribution function", the term "*(probability) distribution*" is frequently used.

When the range $\mathbf{R}(\boldsymbol{\theta})$ of $\boldsymbol{\theta}$, i.e., the set of possible values of $\boldsymbol{\theta}$, is a countable or finite sequence of values $a_1 < a_2 < \ldots$, the distribution of $\boldsymbol{\theta}$ is *discrete*. This means that $F(\theta)$ may be based on a *probability function* $p(\theta)$ which specifies the epistemic probability $p(\boldsymbol{\theta} = a_i)$ attributed to any possible value a_i of $\boldsymbol{\theta}$.

When $\mathbf{R}(\boldsymbol{\theta})$ is an interval of values - as in the above example (iii) - the appropriate distribution $F(\theta)$ on $\boldsymbol{\theta}$ is *continuous*.[5] The derivative of $F(\theta)$ is called the *(probability) density function*, or *pdf*, of $F(\theta)$ and is denoted by $f(\theta)$.

The statistical inferences concerning the value of a *parameter vector* $\boldsymbol{\theta}=(\boldsymbol{\theta}_1,\ldots,\boldsymbol{\theta}_k)$ - where $\boldsymbol{\theta}_1,\ldots,\boldsymbol{\theta}_k$ represent certain quantities characterizing a certain population or process - are called *multivariate inferences*.

Consider, for instance, a *bivariate* parameter vector $\boldsymbol{\theta}=(\boldsymbol{\theta}_1,\boldsymbol{\theta}_2)$. Analogous to the univariate case, the uncertainty about $\boldsymbol{\theta}$ can be represented by a *joint cdf* $F(\theta_1,\theta_2)$. Given a value (a_1,a_2) of $(\boldsymbol{\theta}_1,\boldsymbol{\theta}_2)$, the epistemic probability $p(\boldsymbol{\theta}_1 \leq a_1 \;\&\; \boldsymbol{\theta}_2 \leq a_2)$ that $\boldsymbol{\theta}$ belongs to the 'bidimensional interval' $(\boldsymbol{\theta}_1 \leq a_1, \boldsymbol{\theta}_2 \leq a_2)$ is denoted by $F(a_1,a_2)$:

(3) $$F(a_1,a_2) = p(\boldsymbol{\theta}_1 \leq a_1 \;\&\; \boldsymbol{\theta}_2 \leq a_2)$$

The distinction between *discrete* and *continuous* distributions can also be made in the bivariate case.[6] In particular, a bivariate distribution $F(\theta_1,\theta_2)$ is said to be of continuous type if it is continuous and has a second-order, mixed partial derivative function $f(\theta_1,\theta_2) = \partial^2F/\partial\theta_1\partial\theta_2$ from which $F(\theta_1,\theta_2)$ can be recovered by integration. This derivative will be referred to as the *joint probability density function* - or *joint pdf* - $f(\theta_1,\theta_2)$ on $(\boldsymbol{\theta}_1,\boldsymbol{\theta}_2)$.

From the joint *cdf* $F(\theta_1,\theta_2)$ on $\boldsymbol{\theta}=(\boldsymbol{\theta}_1,\boldsymbol{\theta}_2)$ it is possible to determine the *marginal distributions*, or *marginal cdf*'s, $F(\theta_1)$ and $F(\theta_2)$ on $\boldsymbol{\theta}_1$ and $\boldsymbol{\theta}_2$, and the *conditional distribution* $F(\theta_1/\theta_2)$ expressing the probability distribution on $\boldsymbol{\theta}_1$ given the value θ_2 of $\boldsymbol{\theta}_2$. In particular, from the joint *pdf* $f(\theta_1,\theta_2)$ on a parameter vector $(\boldsymbol{\theta}_1,\boldsymbol{\theta}_2)$, it is possible to determine the *marginal pdf*'s, $f(\theta_1)$ and $f(\theta_2)$, and the *conditional pdf* $f(\theta_1/\theta_2)$.

A possible feature of a *cdf* $F(\theta_1,\theta_2)$ on $(\boldsymbol{\theta}_1,\boldsymbol{\theta}_2)$ is the so-called *independence* which is defined as follows:

(4) $\boldsymbol{\theta}_1$ and $\boldsymbol{\theta}_2$ are *independent* iff $F(\theta_1,\theta_2) = F(\theta_1)F(\theta_2)$.

In particular, given a continuous parameter vector $(\boldsymbol{\theta}_1,\boldsymbol{\theta}_2)$, definition (4) is equivalent to:

(5) $\boldsymbol{\theta}_1$ and $\boldsymbol{\theta}_2$ are *independent* iff $f(\theta_1/\theta_2) = f(\theta_1)$.

Put simply, (5) means that $\boldsymbol{\theta}_1$ and $\boldsymbol{\theta}_2$ are independent iff the prior density function on $\boldsymbol{\theta}_1$ cannot be modified by any subsequent information about $\boldsymbol{\theta}_2$.

All the above definitions can be adapted to a *multivariate cdf* $F(\theta_1,\dots,\theta_k)$ on a parameter vector $\boldsymbol{\theta}=(\boldsymbol{\theta}_1,\dots,\boldsymbol{\theta}_k)$ with $k > 2$. In particular, the notion of *(mutual) independence* of $\boldsymbol{\theta}_1,\dots,\boldsymbol{\theta}_k$, relative to $F(\theta_1,\dots,\theta_k)$ can be defined as follows:

(6) $\boldsymbol{\theta}_1,\dots,\boldsymbol{\theta}_k$ are *(mutually) independent* iff
$$F(\theta_1,\dots,\theta_k) = F(\theta_1)F(\theta_2)\cdots F(\theta_k).$$

Lastly, the *independence of two vectors* $\boldsymbol{\theta}=(\boldsymbol{\theta}_1,\dots,\boldsymbol{\theta}_k)$ and $\boldsymbol{\phi}=(\boldsymbol{\phi}_1,\dots,\boldsymbol{\phi}_h)$ is defined by the following condition:

(7) $F(\theta_1,\dots,\theta_k,\phi_1,\dots,\phi_h) = F(\theta_1,\dots,\theta_k)F(\phi_1,\dots,\phi_h)$

where $F(\theta_1,\dots,\theta_k,\phi_1,\dots,\phi_h)$ is the joint *cdf* on the combined vector $(\boldsymbol{\theta},\boldsymbol{\phi}) = (\boldsymbol{\theta}_1,\dots,\boldsymbol{\theta}_k,\boldsymbol{\phi}_1,\dots,\boldsymbol{\phi}_h)$.

Typical values of a probability distribution

A probability distribution can be described using certain 'typical values' such as mean, variance and moments of various orders which give a general idea of its shape.

The *mean*, or *expected value*, $E(\boldsymbol{\theta})$ of a *cdf* $F(\theta)$ on a parameter $\boldsymbol{\theta}$ is defined as follows:

$$(8) \qquad E(\boldsymbol{\theta}) = \int_{-\infty}^{\infty} \theta \, dF(\theta)$$

where $\int_{-\infty}^{\infty} \theta \, dF(\theta)$ is the Stieltjes integral of θ with respect to $F(\theta)$. In the case of a continuous distribution with *pdf* $f(\theta)$, definition (8) is equivalent to:

$$(9) \qquad E(\boldsymbol{\theta}) = \int_{-\infty}^{\infty} \theta \, f(\theta) \, d\theta$$

For certain purposes one might wish to calculate the mean of certain parameters 'related' to $\boldsymbol{\theta}$. For instance, given a *pdf* $f(\theta)$ on $\boldsymbol{\theta}$, it would be possible to calculate the mean $E(\mathbf{g})$ of the parameter $\mathbf{g} = g(\boldsymbol{\theta})$ - where g: $\mathbf{R}^1 \rightarrow \mathbf{R}^1$ is a continuous function. One could derive the *pdf* $f(g)$ on $\mathbf{g}$ from $f(\theta)$ and, subsequently, apply (9). However a simpler procedure can be used; indeed it can be proved that the value of $E(\mathbf{g})$ is given by the following equality:

$$(10) \qquad E(\mathbf{g}) = \int_{-\infty}^{\infty} g(\theta) \, f(\theta) \, d\theta$$

A similar formula can be applied in the case of a *pdf* $f(\theta) = f(\theta_1,...,\theta_w)$ on $\boldsymbol{\theta}=(\boldsymbol{\theta}_1,...,\boldsymbol{\theta}_w)$. It can be proved that the mean $E(\mathbf{g})$ of the parameter $\mathbf{g} = g(\boldsymbol{\theta}) = g(\boldsymbol{\theta}_1,...,\boldsymbol{\theta}_w)$ - where g: $\mathbf{R}^w \rightarrow \mathbf{R}^1$ is a continuous function - is given by:

$$(11) \qquad E(\mathbf{g}) = \int_{\mathbf{R}^w} g(\theta) f(\theta) \, d\theta$$

where, following La Valle (1970, p. 170), the convention "$d\theta = d\theta_1 \cdot ... \cdot d\theta_w$" is used.

Given a positive integer r, the rth *moment* $\mu_r'(\boldsymbol{\theta})$ and the rth *central moment* $\mu_r(\boldsymbol{\theta})$ of $\boldsymbol{\theta}$ w.r.t. $F(\theta)$ are defined as the expected values of $\boldsymbol{\theta}^r$ and $[\boldsymbol{\theta} - E(\boldsymbol{\theta})]^r$, respectively:

(12) (a) $\mu'_r(\boldsymbol{\theta}) = E(\boldsymbol{\theta}^r)$
(b) $\mu_r(\boldsymbol{\theta}) = E([(\boldsymbol{\theta} - E(\boldsymbol{\theta})]^r)$

The *variance* var($\boldsymbol{\theta}$) of $\boldsymbol{\theta}$ w.r.t. $F(\theta)$ is defined as the second central moment of $\boldsymbol{\theta}$, namely, as the mean of $[\boldsymbol{\theta} - E(\boldsymbol{\theta})]^2$:

(13) $\text{var}(\boldsymbol{\theta}) = E([(\boldsymbol{\theta} - E(\boldsymbol{\theta})]^2)$

Given a multivariate *cdf* $F(\theta_1,..,\theta_k)$ on $\boldsymbol{\theta}=(\boldsymbol{\theta}_1,...,\boldsymbol{\theta}_k)$, the *marginal mean* $E(\boldsymbol{\theta}_i)$ and the *marginal variance* var($\boldsymbol{\theta}_i$) of $\boldsymbol{\theta}_i$ are defined as the mean and the variance of the marginal *cdf* $F(\theta_i)$. The k-tuple $E(\boldsymbol{\theta})=(E(\boldsymbol{\theta}_1),...,E(\boldsymbol{\theta}_k))$ may be called *mean vector* of $F(\theta_1,..,\theta_k)$.

The *covariance* cov($\boldsymbol{\theta}_i, \boldsymbol{\theta}_j$) of $\boldsymbol{\theta}_i$ and $\boldsymbol{\theta}_j$, w.r.t. $F(\theta_1,..,\theta_k)$, is defined as follows:

(14) $\text{cov}(\boldsymbol{\theta}_i, \boldsymbol{\theta}_j) = E([\boldsymbol{\theta}_i - E(\boldsymbol{\theta}_i)][(\boldsymbol{\theta}_j - E(\boldsymbol{\theta}_j)])$

As Lindgren (1976, p. 135) points out, cov($\boldsymbol{\theta}_i, \boldsymbol{\theta}_j$) "purports to measure 'covariation' - to be indicative of the degree to which the variables are concordant". A defect of cov($\boldsymbol{\theta}_i, \boldsymbol{\theta}_j$) is that "it is sensitive to the scale of measurement adopted" (*ibid.*) in measuring $\boldsymbol{\theta}_i$ and $\boldsymbol{\theta}_j$. A measure of covariation that does not have this defect is the *correlation coefficient* $\rho(\boldsymbol{\theta}_i, \boldsymbol{\theta}_j)$, defined as follows:

(15) $$\rho(\boldsymbol{\theta}_i, \boldsymbol{\theta}_j) = \frac{\text{cov}(\boldsymbol{\theta}_i, \boldsymbol{\theta}_j)}{\sqrt{\text{var}(\boldsymbol{\theta}_i)\text{var}(\boldsymbol{\theta}_j)}}$$

The above definitions can also be applied to define the typical values of *conditional* distributions. So, for instance, $E(\boldsymbol{\theta} \mid e)$ would denote the mean of the conditional *cdf* $F(\theta/e)$ where "e" denotes a given empirical evidence.

Finally, notice the properties of typical values as given below.

(16)(a) $E(a\boldsymbol{\theta} + b) = aE(\boldsymbol{\theta}) + b$ (where a and b are real numbers)
(b) $E(\Sigma\, \boldsymbol{\theta}_i) = \Sigma\, E(\boldsymbol{\theta}_i)$

Given the parameter vector $\boldsymbol{\theta}=(\boldsymbol{\theta}_1,...,\boldsymbol{\theta}_w)$, consider the parameter $\mathbf{g} = g(\boldsymbol{\theta}) = g(\boldsymbol{\theta}_1,...,\boldsymbol{\theta}_w)$ where g: $\mathbf{R}^w \rightarrow \mathbf{R}^1$ is a continuous function. Then, if $\boldsymbol{\theta}=(\boldsymbol{\theta}_1,...,\boldsymbol{\theta}_w)$ is partitioned into $\boldsymbol{\theta}_a=(\boldsymbol{\theta}_1,...,\boldsymbol{\theta}_m)$ and $\boldsymbol{\theta}_b=(\boldsymbol{\theta}_{m+1},...,\boldsymbol{\theta}_w)$, formula (11) can be rewritten as follows:

(17) $$E[g(\boldsymbol{\theta}_a,\boldsymbol{\theta}_b)] = \int_{\mathbf{R}^m\mathbf{R}^{w-m}} g(\theta_a,\theta_b)f(\theta_a,\theta_b)\, d\theta_a d\theta_b$$

Given a *pdf* $f(\theta_1,...,\theta_w)$ on $\boldsymbol{\theta}$, the conditional *pdf* $f(\theta_b/\theta_a)$ can be derived. Moreover, given a value θ_a of $\boldsymbol{\theta}_a$, the expectation of the function $g(\theta_a,\cdot)$ of $\boldsymbol{\theta}_b$ w.r.t. the *conditional cdf* $f(\theta_b/\theta_a)$ can be defined. This expectation, which will be referred to as the *conditional expectation of* $g(\theta_a,\boldsymbol{\theta}_b)$ *with respect to* $\boldsymbol{\theta}_b$ given θ_a, is defined as follows (cf. La Valle, 1970, p. 171):

(18) $$E(g(\theta_a,\boldsymbol{\theta}_b) \mid \theta_a) = \int_{\mathbf{R}^{w-m}} g(\theta_a,\theta_b)f(\theta_b/\theta_a)\, d\theta_b$$

It can be proved that the value of $E[g(\boldsymbol{\theta}_a,\boldsymbol{\theta}_b)]$ in (17) is equal to the expected (w.r.t. the marginal *pdf* $f(\theta_a)$) value of $E(g(\theta_a,\boldsymbol{\theta}_b) \mid \theta_a)$ (cf. La Valle, *ibid.*):

(19) $$E[g(\boldsymbol{\theta}_a,\boldsymbol{\theta}_b)] = E[E(g(\boldsymbol{\theta}_a,\boldsymbol{\theta}_b) \mid \boldsymbol{\theta}_a)]$$

It should be noticed that (19) does not hold only in the case of a *pdf* $f(\theta_1,...,\theta_w)$ but for any *cdf* $F(\theta_1,...,\theta_w)$ (cf. Jeffrey, 1971, p. 197).

Some important properties of variance are listed below:

(20) (a) $\mathrm{var}(\boldsymbol{\theta}) \geq 0$ (where $\mathrm{var}(\boldsymbol{\theta}) = 0$ iff the whole probability is concentrated on a single value of $\boldsymbol{\theta}$)

(b) $\mathrm{var}(\boldsymbol{\theta}) = E(\boldsymbol{\theta}^2) - [E(\boldsymbol{\theta})]^2$

(c) $\mathrm{var}(a\boldsymbol{\theta} + b) = a^2\mathrm{var}(\boldsymbol{\theta})$

From (20)(a) and (20)(b) it follows that:

(21) $E(\boldsymbol{\theta}^2) \geq [E(\boldsymbol{\theta})]^2$ (where $E(\boldsymbol{\theta}^2) = [E(\boldsymbol{\theta})]^2$ iff $\mathrm{var}(\boldsymbol{\theta}) = 0$)

Lastly, the following properties of covariance and correlation should be noted:

(22) (a) $\mathrm{cov}(\boldsymbol{\theta}_i,\boldsymbol{\theta}_j) = \mathrm{cov}(\boldsymbol{\theta}_j,\boldsymbol{\theta}_i)$

(b) $\mathrm{cov}(\boldsymbol{\theta}_i,\boldsymbol{\theta}_j) = E(\boldsymbol{\theta}_i\boldsymbol{\theta}_j) - E(\boldsymbol{\theta}_i)E(\boldsymbol{\theta}_j)$

(c) If $\boldsymbol{\theta}_i$ and $\boldsymbol{\theta}_j$ are independent, $\mathrm{cov}(\boldsymbol{\theta}_i,\boldsymbol{\theta}_j) = 0$.

It follows from (15) and (22)(c) that:

(23) If $\boldsymbol{\theta}_i$ and $\boldsymbol{\theta}_j$ are independent, $\rho(\boldsymbol{\theta}_i,\boldsymbol{\theta}_j) = 0$.

Properties (16)-(23) hold for any *cdf* and, hence, also for the typical values $E(\boldsymbol{\theta} \mid e)$, $\mathrm{var}(\boldsymbol{\theta}_i \mid e)$, $\mathrm{cov}(\boldsymbol{\theta}_i, \boldsymbol{\theta}_j \mid e)$, etc. of any *conditional* distribution $F(\theta/e)$ on the parameter vector $\boldsymbol{\theta}$.

A sampling distribution for a (multivariate) Bernoulli process

Consider an experiment $\mathbf{e_n}$ consisting of n trials of a (multivariate) Bernoulli process *Ex* with parameter vector $\mathbf{q}=(\mathbf{q}_1,...,\mathbf{q}_k)$. Any function defined on the possible results e_n of $\mathbf{e_n}$ is called a *statistic* (for $\mathbf{e_n}$). Usually a statistic is a 'descriptive measure' summarizing certain features of the experimental result e_n.

An important statistic is the number $\mathbf{n_i}$ of Q_i -outcomes in $\mathbf{e_n}$. The probability distribution $P_q(n_i)$ on $\mathbf{n_i}$ relative to a given value q of $\mathbf{q}$ - called the *sampling distribution* of $\mathbf{n_i}$ - can be derived from the probability distribution $P_q(e_n)$ (see formula (2.2)). The mean and the variance of $\mathbf{n_i}$ are given by:

$$(24) \qquad E_q(\mathbf{n_i}) = nq_i$$

$$(25) \qquad \mathrm{var}_q(\mathbf{n_i}) = nq_i(1 - q_i)$$

3.3 BAYESIAN STATISTICAL INFERENCES

Following a widely used convention, "(probability distribution) $p(\theta)$" will denote both probability functions and density functions, whenever a given property holds for both.

Using Bayes's theorem, a prior probability distribution $p(\theta)$ on a parameter (vector) $\boldsymbol{\theta}$ can be updated in response to the result y of a certain *experiment* **y**:[7]

$$(26) \qquad p(\theta/y) = \frac{p(y/\theta)p(\theta)}{p(y)}$$

where $p(y)$, in the case of a continuous $p(\theta)$, is given by:

$$(27) \qquad p(y) = \int_{-\infty}^{\infty} p(y/\theta)p(\theta)\, d\theta$$

This formula for the *prior predictive distribution* $p(y)$ can be generalized to a *cdf* $F(\theta)$ as follows:

$$(28) \qquad p(y) = \int_{-\infty}^{\infty} p(y/\theta)\, dF(\theta)$$

Furthermore, given two experiments **y** and **z**, one may calculate the so-called *posterior predictive distribution* $p(y/z)$ where the hypothesis y denotes a possible outcome of **y** and the evidence z a possible outcome of **z**. Suppose that, for each value θ of $\boldsymbol{\theta}$, the experiments **y** and **z** are *conditionally independent* in the sense that $p(y,z/\theta) = p(y/\theta)p(z/\theta)$. Then $p(y/z)$ is given by:[8]

$$(29) \qquad p(y/z) = \frac{\int_{-\infty}^{\infty} p(y/\theta)p(z/\theta)p(\theta)\, d\theta}{p(z)} = \frac{\int_{-\infty}^{\infty} p(y/\theta)p(z/\theta)p(\theta)\, d\theta}{\int_{-\infty}^{\infty} p(z/\theta)p(\theta)\, d\theta}$$

3.4 PROBABILITY DISTRIBUTIONS FOR MULTINOMIAL CONTEXTS

Although a researcher X normally does not know the exact objective probabilities $\mathbf{q}_1,\ldots,\mathbf{q}_k$ governing a given (multivariate) Bernoulli process *Ex* (cf. Chapter 2.3), he might be able to specify a distribution of epistemic probabilities on the possible values of $\mathbf{q}=(\mathbf{q}_1,\ldots,\mathbf{q}_k)$. Given the relation $\mathbf{q}_k = 1 - \mathbf{q}_1 - \cdots - \mathbf{q}_{k-1}$ among the k components of the parameter vector $(\mathbf{q}_1,\ldots,\mathbf{q}_k)$, a k-1-variate *cdf* $F(q_1,\ldots,q_{k-1})$ on the parameter vector $(\mathbf{q}_1,\ldots,\mathbf{q}_{k-1})$ is sufficient to represent the researcher's uncertainty about **q**. For this reason $F(q_1,\ldots,q_{k-1})$ will be called prior (probability) distribution on **q**. In particular, a univariate *cdf* $F(q_1)$ on the parameter $\mathbf{q}_1$ is sufficient to represent the researcher's uncertainty about the probabilities $(\mathbf{q}_1,\mathbf{q}_2)$ governing a given Bernoulli process.

Note that in a Bayesian multinomial context researchers should use the following likelihood function $p(e_n/q) = p(e_n/q_1,\ldots,q_{k-1})$ (cf. formula (2.2)):

$$(30) \qquad p(e_n/q) = p(e_n/q_1,\ldots,q_{k-1}) = \prod_{i=1}^{k} q_i^{n_i}$$

Hence, it follows from formulae (30) and (2.2) that:

(31) (a) $p(Q_i/q) = q_i$
(b) $p(Q_i Q_i/q) = q_i^2$

where "Q_i" and "$Q_i Q_i$" denote the hypothesis that the outcome of the next trial will be Q_i and the hypothesis that the outcomes of the next two trials will be Q_i.

The marginal *cdf*'s of $F(q_1,...,q_{k-1})$ are denoted by $F(q_i)$ and, in particular, the marginal distribution of the 'residual' k-th component $\mathbf{q_k}$ of $(\mathbf{q_1},...,\mathbf{q_k})$ - which can be derived from $F(q_1,...,q_{k-1})$ - is denoted by $F(q_k)$. From the k marginal *cdf*'s $F(q_i)$ one can derive the vector $E(\mathbf{q})$ of the k marginal means $E(\mathbf{q_1}),...,E(\mathbf{q_k})$ which will be called the ***mean vector*** of $F(q_1,...,q_{k-1})$.

The moments of $F(q_1,...,q_{k-1})$ can be interpreted in terms of the *predictive probabilities* attributed to possible sequences of trials of *Ex*. For instance, it follows from (31)(a) and (28) - where in the latter y and θ are replaced with Q_i and q_i, respectively - that the ***marginal mean*** $E(\mathbf{q_i})$ is equal to $p(Q_i)$:

$$(32) \qquad E(\mathbf{q_i}) = \int_0^1 q_i \, dF(q_i) = \int_0^1 p(Q_i/q) \, dF(q_i) = p(Q_i)$$

This means that $E(\mathbf{q_i})$ is identical to the probability $p(Q_i)$ that the next trial of *Ex* is Q_i. Similarly, it follows from (31)(b) and (28) that the second marginal moment $E(\mathbf{q_i^2})$ is equal to the probability $p(Q_i Q_i)$ that the next two trials of *Ex* will be Q_i:

$$(33) \qquad E(\mathbf{q_i^2}) = p(Q_i Q_i)$$

Analogously, it can be proved that:

$$(34) \qquad E(\mathbf{q_i q_j}) = p(Q_i Q_j)$$

Moreover, a *predictive interpretation* of $\text{var}(\mathbf{q_i})$ and $\text{cov}(\mathbf{q_i}, \mathbf{q_j})$ can be provided. In fact, given (31)(b) and (32), the value of $\text{var}(\mathbf{q_i}) = E(\mathbf{q_i^2}) - [E(\mathbf{q_i})]^2$ (cf. (20)(a)) can be expressed by the following formula:

$$(35) \qquad \text{var}(\mathbf{q_i}) = p(Q_i Q_i) - [p(Q_i)]^2$$

while $\text{cov}(\mathbf{q_i}, \mathbf{q_j}) = E(\mathbf{q_i q_j}) - E(\mathbf{q_i})E(\mathbf{q_j})$ (cf. (22)(b)) is given by:[9]

(36) $\text{cov}(\mathbf{q}_i, \mathbf{q}_j) = p(Q_i)[p(Q_j/Q_i) - p(Q_j)]$

Consider the *cdf* $F(\theta_1,..,\theta_k)$ on a given parameter vector $\boldsymbol{\theta}=(\boldsymbol{\theta}_1,...,\boldsymbol{\theta}_k)$. As no standard name exists, the sum $\text{VAR}(\boldsymbol{\theta})$ of the k marginal variances $\text{var}(\boldsymbol{\theta}_i)$ will be referred to as the *total variance* of $F(\theta_1,..,\theta_k)$:

(37) $\text{VAR}(\boldsymbol{\theta}) = \Sigma\, \text{var}(\boldsymbol{\theta}_i)$

In particular, the total variance $\text{VAR}(\mathbf{q})$ of a *cdf* $F(q_1,...,q_{k\text{-}1})$ can be defined as the sum of the k marginal variances $\text{var}(\mathbf{q}_i)$: $\text{VAR}(\mathbf{q}) = \Sigma\text{var}(\mathbf{q}_i)$. It follows from (20)(a) that:

(38) $\text{VAR}(\boldsymbol{\theta}) \geq 0$ (where $\text{VAR}(\boldsymbol{\theta}) = 0$ iff the whole probability is concentrated on a single value of $\boldsymbol{\theta}$)

The following predictive interpretation of $\text{VAR}(\mathbf{q})$ can be derived from the definition of $\text{VAR}(\mathbf{q})$ and (35):

(39) $\text{VAR}(\mathbf{q}) = \Sigma\, (p(Q_iQ_i) - [p(Q_i)]^2)$

where $\Sigma p(Q_iQ_i)$ is the probability that the next two trials of *Ex* belong to the same category.

The above predictive interpretations hold for any *cdf* and, hence, also for the moments $E(\mathbf{q}_i \mid e)$, $\text{var}(\mathbf{q}_i \mid e)$, $\text{cov}(\mathbf{q}_i, \mathbf{q}_j \mid e)$ of any *conditional cdf* $F(q_1,...,q_{k\text{-}1}/e)$.

Notice that the ratio between $\text{var}(\mathbf{q}_i \mid e)$ ($= p(Q_iQ_i/e) - [p(Q_i/e)]^2$: cf. (32)) and $E(\mathbf{q}_i \mid e)$ ($= p(Q_i/e)$: cf. (35)) is given by:[10]

(40) $\text{var}(\mathbf{q}_i \mid e)/E(\mathbf{q}_i \mid e) = p(Q_i/eQ_i) - p(Q_i/e)$

where the quantity on the right side can be interpreted as an '*index of the relevance of Q_i to itself*' (given e).

Notice too that the ratio between $\text{cov}(\mathbf{q}_i, \mathbf{q}_j \mid e)$ ($= p(Q_i/e)[p(Q_j/eQ_i) - p(Q_j/e)]$: cf. (36)) and $E(\mathbf{q}_i \mid e)$ is given by:

(41) $\text{cov}(\mathbf{q}_i, \mathbf{q}_j \mid e)/E(\mathbf{q}_i \mid e) = p(Q_j/eQ_i) - p(Q_j/e)$

where the quantity on the right side can be interpreted as an '*index of the relevance of Q_i to Q_j*' (given e).[11]

Given a value q of $\mathbf{q}$, consider the following two measures: (1) the *uniformity measure* $C(q) = \Sigma q_i^2$ where "C" is in homage to Carnap who proposed Σq_i^2 as a measure of the degree of order of uniformity of a population (for more details see Chapter 8.7, note 10); (2) the *Gini diversity* $G(q) = 1 - \Sigma q_i^2$ (for more details see Chapter 10). W.r.t. these two measures the following typical values of a probability distribution $F(q_1,...,q_{k-1})$ can be considered:

(i) the expected uniformity $E[C(\mathbf{q})]$ and the expected (Gini) diversity $E[G(\mathbf{q})]$ of $\mathbf{q}$;

(ii) the uniformity $C[E(\mathbf{q})] = \Sigma\ [E(\mathbf{q}_i)]^2$ and the (Gini) diversity $G[E(\mathbf{q})] = 1 - \Sigma\ [E(\mathbf{q}_i)]^2$ of the mean vector $E(\mathbf{q})=(E(\mathbf{q}_1),...,E(\mathbf{q}_k))$.

It can be proved that the typical values given above and the total variance $\mathrm{VAR}(\mathbf{q})$ are connected as follows:[12]

(42) $$\mathrm{VAR}(\mathbf{q}) = E[C(\mathbf{q})] - C[E(\mathbf{q})] = G[E(\mathbf{q})] - E[G(\mathbf{q})]$$

Although at first sight one might think that the expected diversity $E[G(\mathbf{q})]$ is identical to the diversity $G[E(\mathbf{q})]$ of the mean vector, it follows from (42) and (38) that these two quantities are usually different from each other and, more specifically, that $E[G(\mathbf{q})]$ does not exceed $G[E(\mathbf{q})]$:

(43) $$0 \leq E[G(\mathbf{q})] \leq G[E(\mathbf{q})]$$

Given a certain mean vector $E(\mathbf{q})$ and the corresponding $G[E(\mathbf{q})]$, if $\mathrm{VAR}(\mathbf{q})$ increases then $E[G(\mathbf{q})]$ decreases so as to approach its minimum value 0, whereas if $\mathrm{VAR}(\mathbf{q})$ decreases then $E[G(\mathbf{q})]$ increases so as to approach its maximum value $G[E(\mathbf{q})]$.

An intuitive illustration of this can be provided by considering two univariate *pdf*'s $f(q_1)$ and $f'(q_1)$ on the binomial parameter $\mathbf{q}=(\mathbf{q}_1,\mathbf{q}_2)$ of a given Bernoulli process *Ex*, where $E(\mathbf{q}_1) = E'(\mathbf{q}_1) = 1/2$ and $\mathrm{var}(\mathbf{q}_1) >> \mathrm{var}'(\mathbf{q}_1)$ (see Figure 2). From $E(\mathbf{q}_1) = E'(\mathbf{q}_1) = 1/2$ it follows that $G[E(\mathbf{q})] = G[E'(\mathbf{q})] = 1/2$, where 1/2 is the maximum value of Gini diversity for a binomial parameter. Moreover, given the equality $\mathrm{var}(\mathbf{q}_2) = \mathrm{var}(\mathbf{q}_1)$ - which always holds for a binomial parameter $\mathbf{q}=(\mathbf{q}_1,\mathbf{q}_2)$ - it follows from $\mathrm{var}(\mathbf{q}_1) >> \mathrm{var}'(\mathbf{q}_1)$ that $\mathrm{VAR}(\mathbf{q}) = 2\mathrm{var}(\mathbf{q}_1) >> 2\mathrm{var}'(\mathbf{q}_1) = \mathrm{VAR}'(\mathbf{q})$. Hence, $\mathrm{VAR}(\mathbf{q}) >> \mathrm{VAR}'(\mathbf{q})$. Given (43) this implies that, notwithstanding the equality $G[E(\mathbf{q})] = G[E'(\mathbf{q})] = 1/2$, the expected diversity $E[G(\mathbf{q})]$ is much lower than the expected diversity $E'[G(\mathbf{q})]$.

Indeed, looking at Figure 2 one sees that the expected diversity $E[G(\mathbf{q})]$ is much smaller than 1/2, since a substantial proportion of the whole probability is concentrated on the two regions near the extremes of the interval [0,1], which are characterized by a very low degree of diversity. On the contrary, $E'[G(\mathbf{q})]$ approaches the maximum value 1/2 (= $G[E'(\mathbf{q})]$) since a substantial proportion of the whole probability is concentrated on the region around the centre, which is characterized by a very high degree of diversity.[13]

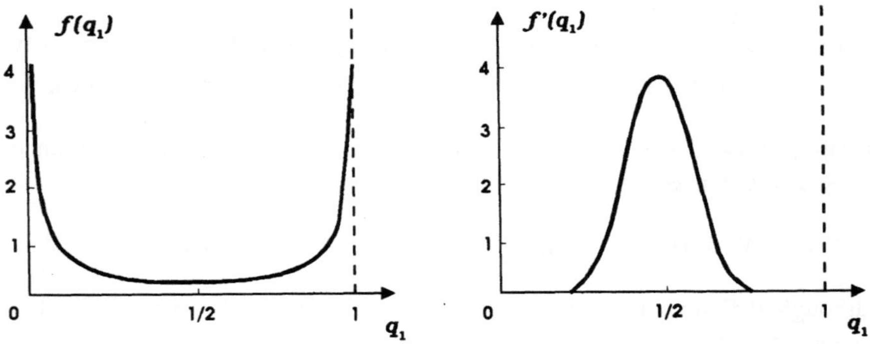

Figure 2

$$E(\mathbf{q}_1) = E'(\mathbf{q}_1) = 1/2 \text{ and } var(\mathbf{q}_1) >> \text{var}'(\mathbf{q}_1)$$
$$\textit{Hence: } G[E(\mathbf{q})] = G[E'(\mathbf{q})] = 1/2 \text{ and } E[G(\mathbf{q})] << E'[G(\mathbf{q})]$$

A predictive interpretation of $E[(C(\mathbf{q})]$ and $E[G(\mathbf{q})]$ can be given thus:[14]

(44) (a) $E[C(\mathbf{q})] = \Sigma\, p(Q_i Q_i)$

(b) $E[G(\mathbf{q})] = 1 - \Sigma\, p(Q_i Q_i)$

where $\Sigma p(Q_i Q_i)$ $(1 - \Sigma p(Q_i Q_i))$ represents the probability that the next two trials of *Ex* belong to the same category (different categories).

3.5 BAYESIAN MULTINOMIAL INFERENCES

The Bayesian statistical inferences made in a multinomial context where a given multivariate Bernoulli process *Ex* is investigated can be referred to as *Bayesian multinomial inferences.*

The prior probability distribution $p(q_1,...,q_{k-1})$ on the parameter vector **q** of a multivariate Bernoulli process *Ex* may be updated on the basis of (the complete description of) a given sequence e_n of *n* trials of *Ex* (cf. Chapter 2.2) using the following formulation of Bayes's theorem (cf. (26)):

$$(45) \qquad p(q_1,...,q_{k-1}/e_n) = \frac{p(e_n/q_1,...,q_{k-1})p(q_1,...,q_{k-1})}{p(e_n)}$$

Recalling that the likelihood function $p(e_n/q_1,...,q_{k-1})$ is given by (30), formula (45) can be rewritten as follows:

$$(46) \qquad p(q_1,...,q_{k-1}/e_n) = \frac{\prod_{i=1}^{k} q_i^{n_i}\, p(q_1,...,q_{k-1})}{p(e_n)}$$

It follows from (27) that, given a continuous $p(q_1,...,q_{k-1})$, the value of $p(e_n)$ in (46) is given by:

$$(47) \qquad p(e_n) = \int_0^1 \int_0^1 \cdots \int_0^1 \prod_{i=1}^{k} q_i^{n_i}\, p(q_1,...,q_{k-1})\, dq_1 \cdot ... \cdot q_{k-1}$$

Using (28) formula (47) can be generalized as follows to an arbitrary *cdf* $F(q)$:

$$(48) \qquad p(e_n) = \int_0^1 \int_0^1 \cdots \int_0^1 \prod_{i=1}^{k} q_i^{n_i}\, dF(q_1 \cdot ... \cdot q_{k-1})$$

Consider the case where a researcher, after performing an experiment $\mathbf{e_n}$ consisting of *n* trials of *Ex*, intends to perform a similar experiment $\mathbf{e_m}$ by performing *m* further trials of *Ex*. Then given a prior $p(q_1,...,q_{k-1})$ on **q** the following predictive distribution $p(e_m/e_n)$ may be derived from (29) by replacing *y* and *z* with e_m and e_n:

$$(49)\qquad p(e_m/e_n) = \frac{\int_0^1\int_0^1 \cdots \int_0^1 p(e_m/q)p(e_n/q)p(q_1,\ldots,q_{k-1})\,dq_1\cdot\ldots\cdot dq_{k-1}}{\int_0^1\int_0^1 \cdots \int_0^1 p(e_n/q)p(q_1,\ldots,q_{k-1})\,dq_1\cdot\ldots\cdot dq_{k-1}}$$

Formula (49) can be generalized to an arbitrary prior *cdf* $F(q_1,\ldots,q_{k-1})$, by replacing "$p(q_1,\ldots,q_{k-1})dq$" with the integral "$dF(q_1,\ldots,q_{k-1})$":

$$(50)\qquad p(e_m/e_n) = \frac{\int_0^1\int_0^1 \cdots \int_0^1 p(e_m/q)p(e_n/q)\,dF(q_1,\ldots,q_{k-1})}{\int_0^1\int_0^1 \cdots \int_0^1 p(e_n/q)\,dF(q_1,\ldots,q_{k-1})}$$

It follows from (50) and (30) that:

$$(51)\qquad p(e_m/e_n) = \frac{\int_0^1\int_0^1 \cdots \int_0^1 q_1^{n_1+m_1}\, q_2^{n_2+m_2} \cdots q_k^{n_k+m_k}\,dF(q_1,\ldots,q_{k-1})}{\int_0^1\int_0^1 \cdots \int_0^1 q_1^{n_1}\, q_2^{n_2} \cdots q_k^{n_k}\,dF(q_1,\ldots,q_{k-1})}$$

Notice that the denominator of (51) represents the prior predictive probability $p(e_n)$ (cf. formula (48)).

The *special value*, or *singular* predictive probability, $p(Q_i/e_n)$ is simply a particular case of (51):

$$(52)\qquad p(Q_i/e_n) = \frac{\int_0^1\int_0^1 \cdots \int_0^1 q_1^{n_1}\, q_2^{n_2} \cdots q_i^{n_i+1} \cdots q_k^{n_k}\,dF(q_1,\ldots,q_{k-1})}{\int_0^1\int_0^1 \cdots \int_0^1 q_1^{n_1}\, q_2^{n_2} \cdots q_i^{n_i} \cdots q_k^{n_k}\,dF(q_1,\ldots,q_{k-1})}$$

Given that an inductive method (for *Ex*) is virtually identical to a prior predictive distribution on *Ex* or, equivalently, to a given class of special values (see Chapter 2.1), it follows from (48) and (52) that any *cdf* $F(q_1,\ldots,q_{k-1})$ on the parameter (vector) **q** of a (multivariate) Bernoulli process

Ex determines a unique inductive method for *Ex*. For instance, the uniform distribution on the parameter $\mathbf{q}_1$ - which attributes equal density $f(q_1) = 1$ to all the values of $\mathbf{q}_1$ - determines Laplace's *rule of succession*, i.e., the inductive method characterized by the special values: $p(Q_i/e_n) = (n_i + 1)/(n + 2)$.[15]

Lastly, it can be proved that:

(53) The inductive method *I* derived from $F(q_1,...,q_{k-1})$ satisfies:
(a) the requirement of exchangeability (Exc);
(b) the requirement of non-negative relevance (NNR).[16]

CHAPTER 4

BAYESIAN POINT ESTIMATION, VERISIMILITUDE, AND IMMODESTY

The principal objective of this chapter is to prove that epistemic probability distributions satisfy the so-called property of *immodesty* which plays a very important role in the analysis of EPO given herein (cf. Chapter 8.6).

In Section 1 the *mean estimator*, which takes the mean (vector) of the posterior probability distribution as the point estimate of the parameter (vector) $\boldsymbol{\theta}$, is considered and a *verisimilitude interpretation* of this estimator is given.

With reference to this interpretation the notion of *immodesty* is illustrated and the immodesty of epistemic probability distributions is proved (Section 2). In addition some implications of this property relative to the probability distributions on the parameter (vector) $\mathbf{q}$ of a (multivariate) Bernoulli process are considered.

Lastly, in Section 3 a brief outline of the basic issues of the verisimilitude theory is given.

4.1 BAYESIAN POINT ESTIMATION AND VERISIMILITUDE

A *point estimator* is an inductive procedure used for making *point estimates* of the unknown value of a given quantity, on the basis of certain possible experimental data. More precisely, given a parameter (vector) $\boldsymbol{\theta}$ and an experiment $\mathbf{y}$, a point estimator T: $\mathbf{y} \rightarrow \mathbf{R}(\boldsymbol{\theta})$ is a real-valued function which assigns to each possible result y of $\mathbf{y}$ a point estimate $T(\boldsymbol{\theta} \mid y) \in \mathbf{R}(\boldsymbol{\theta})$ - where $\mathbf{R}(\boldsymbol{\theta})$ is the range of the possible values of $\boldsymbol{\theta}$.

A commonly used Bayesian point estimator - which will be referred to as *mean estimator* or *M-estimator E* - takes the *mean* (*vector*) $E(\boldsymbol{\theta} \mid y)$ of the posterior probability distribution $p(\theta/y)$ as the point estimate of $\boldsymbol{\theta}$.

As shown below the M-estimator E may be seen as a particular application of the Bayes's principle which recommends the action which minimizes the expected loss. A *loss function* $L(a,\theta)$ denotes the loss resulting from the adoption of the point estimate a when $\boldsymbol{\theta} = \theta$. In particular, given a parameter $\boldsymbol{\theta}$, the *quadratic loss* $L_2(a,\theta)$ is defined as follows:[1]

(1) $L_2(a,\theta) = (\theta - a)^2$

Similarly, given the estimate $a=(a_1,...,a_k)$ of a parameter vector $\boldsymbol{\theta}=(\boldsymbol{\theta}_1,...,\boldsymbol{\theta}_k)$, the quadratic loss $L_2(a,\theta)$ is defined thus:

(2) $L_2(a,\theta) = \Sigma (\theta_i - a_i)^2$

Some standard results concerning the expected (w.r.t. $p(\theta/y)$) quadratic loss $E(L_2(a,\boldsymbol{\theta}) \mid y)$ of the estimate a of a parameter (vector) $\boldsymbol{\theta}$ are given by the following formulae (3)-(4):

(3) Given a parameter $\boldsymbol{\theta}$, $E(L_2(a,\boldsymbol{\theta}) \mid y) = \text{var}(\boldsymbol{\theta} \mid y) + (E(\boldsymbol{\theta} \mid y) - a)^2$.

Given a parameter vector $\boldsymbol{\theta}$,

$$E(L_2(a,\boldsymbol{\theta}) \mid y) = \text{VAR}(\boldsymbol{\theta} \mid y) + \Sigma (E(\boldsymbol{\theta}_i \mid y) - a_i)^2.$$

As an immediate consequence of (3):

(4) $E(L_2(a,\boldsymbol{\theta}) \mid y)$ is minimized by $a = E(\boldsymbol{\theta} \mid y)$.

This means that the *M*-estimator *E* is the appropriate strategy for minimizing the expected quadratic loss.

In the statistical literature the losses employed in the 'justification' of *M*-estimators and other Bayesian estimators are usually described as pecuniary or, more generally, pragmatic. However, in the epistemological *verisimilitude theory* a different view concerning the *nature* of such losses has recently been expounded. According to this view, such losses should be interpreted - at least in *pure* science - as *cognitive losses* representing the 'distance of an estimate *a* from the truth θ'.[2] Accordingly, *M*-estimators should be considered as the appropriate strategies for *minimizing the expected quadratic distance* (of our estimates) *from the truth* (the true value of $\boldsymbol{\theta}$). The interpretation of quadratic losses as 'quadratic distances' will be emphasized by putting $D_2(a,\theta)$ in place of $L_2(a,\theta)$.[3]

It is often argued that "the uncertainty about the quantity of interest ... might be measured by the variance" (Winkler, 1972, p. 155). The *verisimilitude interpretation* of quadratic losses ($L_2(a,\theta) = D_2(a,\theta)$) allows a more precise formulation of this claim. Indeed, given this interpretation, it follows from (3) that the variance $\text{var}(\boldsymbol{\theta} \mid y)$ of the posterior probability distribution $p(\theta/y)$ on a given parameter $\boldsymbol{\theta}$ is equal to *the expected quadratic distance of* the expected value $E(\boldsymbol{\theta} \mid y))$ *from the true value of* $\boldsymbol{\theta}$: $\text{var}(\boldsymbol{\theta} \mid y) = E(D_2(E(\boldsymbol{\theta} \mid y),\boldsymbol{\theta}) \mid y)$. Analogously, as far as the total variance $\text{VAR}(\boldsymbol{\theta} \mid y)$ of the posterior probability distribution $p(\theta/y)$ on a parameter vector $\boldsymbol{\theta}$ is concerned, it follows from (3) that $\text{VAR}(\boldsymbol{\theta} \mid y) = E(D_2(E(\boldsymbol{\theta} \mid y),\boldsymbol{\theta}) \mid y)$. Obviously these equalities also apply to the case where y is a tautological evidence. In particular, the following equality concerning the total variance of a given prior distribution $p(\theta)$ on a parameter vector $\boldsymbol{\theta}$ will be used later in this book:

$$(5) \qquad \text{VAR}(\boldsymbol{\theta}) = E[D_2(E(\boldsymbol{\theta}),\boldsymbol{\theta})]$$

4.2 THE PRINCIPLE OF IMMODESTY

Immodest inductive methods and the related *requirement of immodesty* are informally illustrated by Lewis (1971, p. 54) as follows:

> Inductive methods can be used to estimate the accuracies of inductive methods. Call a method *immodest* if it estimates that it is at least as accurate as any of its rivals. It would be unreasonable to adopt any but an immodest method.

After observing that an inductive method **C** is used to make estimates of certain magnitudes, where such estimates are equated to the *expected values* of those magnitudes calculated on the basis of **C**, Lewis (*ibid.*, pp. 54-55) explains the concept of immodest inductive method as follows:

> You should hope to give your trust to an inductive method C that will give you accurate estimates ...
>
> But you cannot just pick the most accurate method - not unless you already know the actual values of the magnitude you wish to estimate in which case you do not need to estimate them. The best you can do is pick up the inductive method with the highest *estimated* accuracy, just as you might bet on the horse with the highest estimated speed.
>
> The trouble is that you need an inductive method to estimate anything, even to estimate the accuracy of various inductive methods. And your selection of a method with the highest estimated accuracy will come out differently depending on which method you use to make the estimate. It is as if *Consumer Reports, Consumer Bulletin*, etc., each published rankings of the consumers' magazines, as they do of other products. You would have to know which one to read in order to find out which one to read. Let us say that an inductive method C *recommends* an inductive method C' if the C-mean estimate of the accuracy of C' is not exceeded by the C-mean estimate of the accuracy of any rival method. An inductive method might or might not recommend itself. If it does, let us call it *immodest*. When asked which method has the best estimated accuracy, the immodest method answers: "I have".

Lewis (*ibid.*, p. 56) argues most convincingly that immodesty is a necessary requirement for adopting an inductive method:

.... would *non*-immodesty give you any good reason *not* to trust an inductive method? Indeed it would. Suppose you did trust some non-immodest method. By definition, it estimates some competing method to be more accurate than itself. So if you really did trust your original method, you should take its advice and transfer your trust to one of the competing methods it recommends. It is as if *Consumer Bulletin* were to advise you that *Consumer Reports* was a best buy whereas *Consumer Bulletin* itself was not acceptable; you could not possibly trust *Consumer Bulletin* completely thereafter.[4]

On the other hand, Lewis (1971, p. 56) recognizes that "immodesty is a necessary but not sufficient condition of adequacy for inductive methods". Accordingly, if a given class of inductive methods includes several immodest ones, then they "must compete among themselves on other grounds".

Regarding Carnap's inductive methods (1952) - which are a particular class of exchangeable inductive methods (cf. Chapter 1.3 and 6.1) - Lewis *mistakenly* believes that the requirement of immodesty can restrict the choice to a uniquely admissible immodest method.[5] In the next subsection it will be shown - contrary to Lewis's assumption - that *any* exchangeable inductive method is immodest. Therefore immodesty *as such* cannot impose any genuine restriction on exchangeable inductive methods and, in particular, it cannot impose any restriction on Carnap's methods. On the other hand, it will be seen that immodesty plays an important - although more indirect - role in the present contextual approach to the epistemological problem of optimality (cf. Chapter 8.6).

Immodest probability distributions

Suppose that a researcher investigating a given parameter (vector) $\boldsymbol{\theta}$ intends to perform an experiment **y**. Remember that, for any possible result y of **y** the expected (w.r.t. the posterior probability distribution $p(\theta/y)$) quadratic distance $E(D_2(a,\boldsymbol{\theta}) \mid y)$ is minimized by $a = E(\boldsymbol{\theta} \mid y)$ (see (4)). In other words, for any possible result y of **y**, the *M*-estimator E is the appropriate strategy for minimizing the expected quadratic distance from the truth (the true value of $\boldsymbol{\theta}$).

Now consider the moment where the experiment **y** has still to be performed and, therefore, the posterior distribution $p(\theta/y)$ has not yet been obtained. Then the researcher has at his disposal only a given prior probability distribution $p(\theta)$ on $\boldsymbol{\theta}$, a given likelihood function $p(y/\theta)$, and the derived joint distribution $p(\theta,y)$ $(= p(\theta)p(y/\theta))$. Now a prior distribution $p(\theta)$ is said to satisfy the *principle of immodesty* if, for any experiment **y** and any likelihood function $p(y/\theta)$, the expected (w.r.t. $p(\theta,y)$) quadratic distance between the *M*-estimates of $\boldsymbol{\theta}$ made by the $p(\theta)$-based *M*-estimator *E* and the truth is not higher than the quadratic distance between the estimates of $\boldsymbol{\theta}$ made by any other estimator and the truth. In order to *prove* that any prior distribution on $\boldsymbol{\theta}$ is immodest, a more rigorous definition of *immodest prior distribution* is needed.

Given the result *y* of an experiment **y**, let $E(\boldsymbol{\theta} \mid y)$ and $T(\boldsymbol{\theta} \mid y)$ be the point estimates of $\boldsymbol{\theta}$ made by the *M*-estimator *E* and an arbitrary point estimator *T*. Moreover, given the statistics $E(\boldsymbol{\theta} \mid \mathbf{y})$ and $T(\boldsymbol{\theta} \mid \mathbf{y})$, let $E[D_2(E(\boldsymbol{\theta} \mid \mathbf{y}),\boldsymbol{\theta})]$ denote the expected (w.r.t. $p(\theta,y)$) distance between $E(\boldsymbol{\theta} \mid \mathbf{y})$ and $\boldsymbol{\theta}$, and $E[D_2(T(\boldsymbol{\theta} \mid \mathbf{y}),\boldsymbol{\theta})]$ the expected (w.r.t. $p(\theta,y)$) distance between $T(\boldsymbol{\theta} \mid \mathbf{y})$ and $\boldsymbol{\theta}$.

Immodest priors can now be defined as follows:

(6) A prior distribution $p(\theta)$ on the parameter (vector) $\boldsymbol{\theta}$ is *immodest* iff, for any experiment **y**, any likelihood function $p(y/\theta)$ and any estimator *T*:

$$E[D_2(E(\boldsymbol{\theta} \mid \mathbf{y}),\boldsymbol{\theta})] \leq E[D_2(T(\boldsymbol{\theta} \mid \mathbf{y}),\boldsymbol{\theta})]$$

In other words $p(\theta)$ is immodest iff $E[D_2(T(\boldsymbol{\theta} \mid \mathbf{y}),\boldsymbol{\theta})]$ is minimized by the estimator $T = E$.

In particular, it should be pointed out that an immodest prior $p(\theta)$ is '*immodest w.r.t. all other prior distributions*' $p'(\theta)$ *on* $\boldsymbol{\theta}$ in the sense that an immodest prior $p(\theta)$ 'estimates' that the estimates of $\boldsymbol{\theta}$ made by the $p(\theta)$-based *M*-estimator *E* will be no more distant from the truth than the estimates of $\boldsymbol{\theta}$ made by any other estimator *T*, *including* any other $p'(\theta)$-based *M*-estimator *E'*.[6]

Before demonstrating the universality of immodesty, the following lemma must be proved:

(7) Given a parameter $\boldsymbol{\theta}$,

$$E[D_2(T(\boldsymbol{\theta} \mid \mathbf{y}),\boldsymbol{\theta})] = E[\mathrm{var}(\boldsymbol{\theta} \mid \mathbf{y})] + E(D_2[E(\boldsymbol{\theta} \mid \mathbf{y}) - T(\boldsymbol{\theta} \mid \mathbf{y})]);$$

Given a parameter vector $\boldsymbol{\theta}$,

$$E[D_2(T(\boldsymbol{\theta} \mid \mathbf{y}),\boldsymbol{\theta})] = E[\mathrm{VAR}(\boldsymbol{\theta} \mid \mathbf{y})] + E(D_2[E(\boldsymbol{\theta} \mid \mathbf{y}) - T(\boldsymbol{\theta} \mid \mathbf{y})]).$$

where (i) $E[D_2(T(\boldsymbol{\theta} \mid \mathbf{y}),\boldsymbol{\theta})]$ is the expected (w.r.t. $p(\theta,y)$) quadratic distance between $T(\boldsymbol{\theta} \mid \mathbf{y})$ and $\boldsymbol{\theta}$; (ii) $E[\mathrm{var}(\boldsymbol{\theta} \mid \mathbf{y})]$, $E[\mathrm{VAR}(\boldsymbol{\theta} \mid \mathbf{y})]$ and

$E(D_2[E(\boldsymbol{\theta} \mid \mathrm{y}) - T(\boldsymbol{\theta} \mid \mathrm{y})])$ are the expected (w.r.t. $p(y)$) values of $\mathrm{var}(\boldsymbol{\theta} \mid \mathrm{y})$, $\mathrm{VAR}(\boldsymbol{\theta} \mid \mathrm{y})$ and $D_2[E(\boldsymbol{\theta} \mid \mathrm{y}) - T(\boldsymbol{\theta} \mid \mathrm{y})]$.[7]

Proof: Given a parameter $\boldsymbol{\theta}$, let $E(D_2(T(\boldsymbol{\theta} \mid y),\boldsymbol{\theta}) \mid y)$ be the expected (w.r.t. $p(\theta/y)$) value of $D_2(T(\boldsymbol{\theta} \mid y),\boldsymbol{\theta})$. Then the expected (w.r.t. $p(y)$) value of $E(D_2(T(\boldsymbol{\theta} \mid \mathrm{y}),\boldsymbol{\theta}) \mid \mathrm{y})$ will be referred to as $E[E(D_2(T(\boldsymbol{\theta} \mid \mathrm{y}),\boldsymbol{\theta}) \mid \mathrm{y})]$. It follows from formula (3.19) that $E[E(D_2[T(\boldsymbol{\theta} \mid \mathrm{y}),\boldsymbol{\theta}] \mid \mathrm{y})]$ is identical to $E[D_2(T(\boldsymbol{\theta} \mid \mathrm{y}),\boldsymbol{\theta})]$:

(a) $$E[D_2(T(\boldsymbol{\theta} \mid \mathrm{y}),\boldsymbol{\theta})] = E[E(D_2(T(\boldsymbol{\theta} \mid \mathrm{y}),\boldsymbol{\theta}) \mid \mathrm{y})]$$

In addition, it follows from (3), by replacing a with $T(\boldsymbol{\theta} \mid y)$, that the expected (w.r.t. $p(\theta/y)$) quadratic loss $E(L_2(T(\boldsymbol{\theta} \mid y),\boldsymbol{\theta}) \mid y)$ is given by:

(b) $$E(L_2(T(\boldsymbol{\theta} \mid y),\boldsymbol{\theta}) \mid y) = \mathrm{var}(\boldsymbol{\theta} \mid y) + [E(\boldsymbol{\theta} \mid y) - T(\boldsymbol{\theta} \mid y)]^2$$

Equality (b) can be restated as follows:

(c) $$E(D_2(T(\boldsymbol{\theta} \mid y),\boldsymbol{\theta}) \mid y) = \mathrm{var}(\boldsymbol{\theta} \mid y) + D_2[E(\boldsymbol{\theta} \mid y) - T(\boldsymbol{\theta} \mid y)]$$

From (c) and (3.16.a) it follows that the expected (w.r.t. $p(y)$) value $E[E(D_2(T(\boldsymbol{\theta} \mid \mathrm{y}),\boldsymbol{\theta}) \mid \mathrm{y})]$ of $E(D_2(T(\boldsymbol{\theta} \mid \mathrm{y}),\boldsymbol{\theta}) \mid \mathrm{y})$ is:

(d) $$\begin{aligned} E[E(D_2(T(\boldsymbol{\theta} \mid \mathrm{y}),\boldsymbol{\theta}) \mid \mathrm{y})] &= E(\mathrm{var}(\boldsymbol{\theta} \mid \mathrm{y}) + D_2[E(\boldsymbol{\theta} \mid \mathrm{y}) - T(\boldsymbol{\theta} \mid \mathrm{y})]) \\ &= E[\mathrm{var}(\boldsymbol{\theta} \mid \mathrm{y})] + E(D_2[E(\boldsymbol{\theta} \mid \mathrm{y}) - T(\boldsymbol{\theta} \mid \mathrm{y})]) \end{aligned}$$

The first part of (7) is an immediate consequence of (a) and (d). The second part of (7) can be proved similarly, *Q.E.D.*

Using (7) it can be proved that:

(8) Any prior distribution $p(\theta)$ on a parameter (vector) $\boldsymbol{\theta}$ is *immodest*.

Proof: Given a prior distribution $p(\theta)$ on a parameter $\boldsymbol{\theta}$, consider the equality $E[D_2(T(\boldsymbol{\theta} \mid \mathrm{y}),\boldsymbol{\theta})] = E[\mathrm{var}(\boldsymbol{\theta} \mid \mathrm{y})] + E(D_2[E(\boldsymbol{\theta} \mid \mathrm{y}) - T(\boldsymbol{\theta} \mid \mathrm{y})])$ in (7). Since $E[\mathrm{var}(\boldsymbol{\theta} \mid \mathrm{y})]$ is constant and since $E(D_2[E(\boldsymbol{\theta} \mid \mathrm{y}) - T(\boldsymbol{\theta} \mid \mathrm{y})])$ is necessarily non-negative and equal to 0 only in the case where, for any y, $T(\boldsymbol{\theta} \mid y) = E(\boldsymbol{\theta} \mid y)$, then the expected distance $E[D_2(T(\boldsymbol{\theta} \mid \mathrm{y}),\boldsymbol{\theta})]$ is minimized by the estimator $T = E$. This signifies that $p(\theta)$ is immodest, given the definition (6) of immodesty.

An analogous proof can be provided for a parameter vector $\boldsymbol{\theta}$, *Q.E.D.*

Definition (6) and theorem (8) refer to an arbitrary prior distribution $p(\theta)$ where $p(\theta)$ may be a probability function or a density function (see the first lines of Chapter 3.3). However, it is clear that both definition (6) and theorem (8) can be generalized to apply to any prior *cdf* $F(\theta)$ on $\boldsymbol{\theta}$. In particular, theorem (8) can be extended as follows:

(9) Any prior *cdf* $F(\theta)$ on a parameter (vector) $\boldsymbol{\theta}$ is *immodest.*[8]

For our purposes the following particular consequence of (9) should be considered:

(10) Any prior *cdf* $F(q_1,...,q_{k-1})$ on the parameter (vector) $\mathbf{q}$ of a (multivariate) Bernoulli process is *immodest.*

Since any exchangeable inductive method is 'equivalent' to a corresponding prior *cdf* $F(q_1,...,q_{k-1})$ (see Chapters 1.3 and 5.1), it follows from (10) that:

(11) Any exchangeable inductive method is immodest.

The theorems (10)-(11) generalize some results obtained by Spielman (1972) and Horwich (1982).[9]

Finally, consider the following theorem which will be of fundamental importance in the analysis of EPO (see Chapter 8.6). Given two distributions $F(q_1,...,q_{k-1})$ and $F'(q_1,...,q_{k-1})$ on the parameter (vector) $\mathbf{q}$ - and a likelihood function $p(e_n/q)$ - let E and E' be the corresponding M-estimators. Then, given an experiment $\mathbf{e}_n$, where the results are the possible sequences e_n of n trials of *Ex*, it follows from (6) and (10) that:

(12) $E[D_2(E(\mathbf{q} \mid \mathbf{e}_n),\mathbf{q})] \leq E[D_2(E'(\mathbf{q} \mid \mathbf{e}_n),\mathbf{q})]$

where $E[D_2(E(\mathbf{q} \mid \mathbf{e}_n),\mathbf{q})]$ and $E[D_2(E'(\mathbf{q} \mid \mathbf{e}_n),\mathbf{q})]$ are the expected (w.r.t. the joint probability distribution on $(\mathbf{q},\mathbf{e}_n)$ derived from $F(q_1,...,q_{k-1})$ and $p(e_n/q)$) values of $D_2(E(\mathbf{q} \mid \mathbf{e}_n),\mathbf{q})$ and $D_2(E'(\mathbf{q} \mid \mathbf{e}_n),\mathbf{q})$, respectively.

4.3 THE BASIC ISSUES OF THE VERISIMILITUDE THEORY

The verisimilitude interpretation of *M*-estimators (see Section 1) is a small part of a methodological programme of research dealing with two related problem areas: (1) the *logical* problem of verisimilitude and (2) the *epistemic* problem of verisimilitude. The first problem area involves questions such as: what does it mean to say that a theory is closer to the truth than another or that a given theory has a certain degree of verisimilitude? The second problem area involves questions such as: how can one rationally conjecture, on some evidence, that one theory is closer to the truth than another? how can one conjecturally evaluate the degree of verisimilitude of a given theory?

The first systematic attempt to define a formal concept of verisimilitude, or truthlikeness, for scientific theories was made at the beginning of the sixties by Popper. A vivid account of the impact of Popper's comparative notion of verisimilitude on the epistemological community is provided by Kuipers (1987, p. 1):

> When Karl Popper published in 1963 his definition of closer-to-the truth this was an important intellectual event, but not a shocking one. Everybody could react by saying that the definition was as it should be, and even that it could have been expected. For plausible the definition was indeed: a theory is closer to the truth than another one if the first has more true and less false consequences than the second.
>
> About then years later the 1963 event became shocking with retrospective effect when David Miller and Pavel Tichý independently proved that a false theory, that is a theory with at least one false consequence, could according to Popper's definition never be closer to the truth than another one.
>
> With this proof they demolished the definition, for it could not do justice to the presupposed nature of most of the culminating-points in the history of science: new theories, such as Einstein's theory, though presumably false, are more successful than their predecessors, such as Newton's theory, just because they are closer to the truth, that is, closer to the unknown true theory about the subject matter.
>
> Miller and Tichý unchained with their proof in the beginning ... mainly sceptical remarks like "only the intuitive idea is important, fortunately", "it shows that one can't solve philosophical problems by formal means" and last but not least "it is the punishment for speaking freely about *the* truth".

On the other hand, after some time the realization that Popper's definition was not almost obviously correct but (indeed surprisingly) was mistaken, stimulated the attention of an increasing number of epistemologists towards verisimilitude.[10]

Most of them abandoned the *consequence approach* proposed by Popper and explored alternative approaches to the explication of verisimilitude.[11] As a result of almost twenty years on research there are now several formal approaches to the logical problem of verisimilitude.

The most widely supported approach is the so-called *similarity approach* developed by Tichý, Oddie, Tuomela, Niiniluoto and others. The basic idea of this approach is that the verisimilitude (distance from the truth) of a hypothesis *h* depends on the similarities between the states of affairs allowed by *h* and the true state of the world. The similarity approach is primarily syntactic since - as Kuipers (1987a, p. 2) remarks - it "centers around the syntactic similarities and dissimilarities between the relevant potential answers to the [cognitive] problem" under consideration where a cognitive problem is characterized by a specific language and 'the truth' is identified with the most informative true statement in that language.

A possible alternative to the similarity approach to verisimilitude is the *semantic approach* developed by Miller (1978) and Kuipers (1982, 1987b). They start from the assumption of a previously given set of conceptual possibilities or, more precisely, of a given set of conceptually possible structures, which may or may not be explicitly based on some given language. This means that within the semantic approach "the appropriate category for defining truthlikeness is not the category of sentences but of (sets of) models of sentences or, more generally, structures and sets of structures" (Kuipers, 1987a, p. 4). For instance, within Kuipers's formulation of the semantic approach, a distinction between the so-called descriptive truth and theoretical truth is drawn, where the *descriptive truth* is given by one structure representing the actual world, and the *theoretical truth* is defined as the set of structures representing the set of physically possible worlds (or set of empirical possibilities).[12]

Different methodological rules have been suggested for the conjectural evaluation of the verisimilitude of a given hypothesis - and the conjectural comparison of the verisimilitude of two hypotheses - on the ground of the available evidence.

For instance, according to the Bayesian approach to the epistemic problem of verisimilitude, an appropriate procedure consists of *estimating* the verisimilitude (distance from the truth) of a hypothesis on evidence by calculating its expected degree of verisimilitude (distance from the truth) relative to an epistemic probability measure.[13] Within the Bayesian approach other procedures for evaluating the verisimilitude of hypotheses can be explored. Suppose, for instance, that the maximum verisimilitude is 1. Then the evaluation of a given hypothesis *h* may be based on the "degree of *probable verisimilitude*" of *h* which is defined as the probability, on a given evidence *e*, that the degree of verisimilitude of *h* exceeds a certain threshold $1 - \varepsilon$ (see Niiniluoto 1987, p. 278; 1989, pp. 237-239).[14]

A limitation of the Bayesian approach "is that it is almost inevitably assuming a quantitative truthlikeness" (Kuipers, 1987b, p. 95). With reference to this Kuipers (*ibid.*) maintains that an adequate rule of choice between two rival hypotheses "should also, and primarily, deal with the comparative case, without necessarily presupposing an explication of quantitative truthlikeness". In several papers Kuipers develops a non-Bayesian approach to the epistemic problem of verisimilitude where a number of methodological applications of his comparative notions of truthlikeness are explored.[15]

PART II

DE FINETTI'S THEOREM, GC-SYSTEMS, AND DIRICHLET DISTRIBUTIONS

CHAPTER 5

EXCHANGEABLE INDUCTIVE METHODS, BAYESIAN STATISTICS, AND CONVERGENCE TOWARDS THE TRUTH

In this chapter the so-called *de Finetti's representation theorem* (*DFRT*) is used to elucidate some conceptual relationships between TIP and the analysis of multinomial inferences in BS.

In Section 1 some implications of DFRT are noted. For instance, there is a one-to-one correspondence between the set of *exchangeable* inductive methods and the set of prior distributions on the parameter (vector) $\mathbf{q}$ of a (multivariate) Bernoulli process.

In Section 2 some problems regarding the convergence of opinion in scientific research are considered. In particular, it is shown that exchangeable inductive methods which satisfy the requirement of adequacy (Reich) allow scientists to learn from experience, not only in the 'epistemic sense' that scientists using different inductive methods are led to a considerable convergence of opinion, but also in the 'realist sense' that the predictive probabilities derived from such methods tend, *in the long run*, to approach the truth.

5.1 DE FINETTI'S REPRESENTATION THEOREM

At the end of Chapter 3 it was pointed out that any *cdf* $F(q_1,...,q_{k-1})$ on the parameter (vector) $\mathbf{q}$ of a (multivariate) Bernoulli process *Ex* determines a unique inductive method for *Ex* and that, in addition, (cf. (3.53)):

(1) The inductive method *I* derived from $F(q_1,...,q_{k-1})$ satisfies the requirement of exchangeability (Exc).

It may be asked whether the reverse also holds true, i.e., whether an exchangeable inductive method *I* for *Ex* 'determines' a (unique) probability distribution $F(q_1,...,q_{k-1})$.

Rather surprisingly a positive answer to this question is provided by DFRT, that is:

(2) For any (predictive distribution p_I specified by an) *exchangeable* inductive method I for *Ex*,

(a) a *cdf* $F_I(q_1,...,q_{k-1})$ *does exist* such that, for any sequence e_n of trials of *Ex*:

$$p_I(e_n) = \int_0^1 \int_0^1 \cdots \int_0^1 \prod_{i=1}^{k} q_i^{n_i} \, dF_I(q_1, \ldots, q_{k-1})$$

(b) $F_I(q_1,...,q_{k-1})$ is *unique*.[1]

Put simply, DFRT asserts that any exchangeable inductive method I for *Ex* can be derived from, or 'represented' by, one and only one *cdf* $F_I(q_1,...,q_{k-1})$. Note that $F_I(q_1,...,q_{k-1})$ is, or at least behaves as, a probability distribution on the parameter (vector) $\mathbf{q}$ of a (multivariate) Bernoulli process.[2]

DFRT can be used to clarify the conceptual relationships between TIP and the analysis of multinomial inferences made in BS. Note, for instance, that, given (1) and (2) then:

(3) There is a one-to-one correspondence, or 'equivalence', between the set of exchangeable inductive methods and the set of the probability distributions on the parameter (vector) $\mathbf{q}$ of a (multivariate) Bernoulli process.[3]

More specifically, DFRT can be used to prove the equivalence between a given class of exchangeable inductive methods and a corresponding class of probability distributions on $\mathbf{q}$. An example of this kind is given by the equivalence between GC-systems and Dirichlet distributions which can be proved by using DFRT (see Chapter 6.3).

Lastly, DFRT could be applied to identify the conceptual relations existing between:

(i) certain *predictive requirements* of adequacy concerning special values and other predictive probabilities as employed in TIP (see Chapter 2.1), and

(ii) certain *parametric requirements* concerning the form of a probability distribution on the parameter (vector) $\mathbf{q}$ of a (multivariate) Bernoulli process as considered in BS (see Chapter 6.6).

For instance, DFRT could be used to prove that a given predictive requirement *A* is *equivalent* to a parametric requirement *B*, in the sense that an exchangeable inductive method satisfies *A* iff the corresponding probability distribution on **q** satisfies *B*.

Consider, for example, the following *'no gap'* parametric requirement (NG):

(NG) Given the *cdf* $F(q_1)$ on the parameter $\mathbf{q}_1$ of a Bernoulli process, $p(H) > 0$ for any non-degenerate interval *H* in [0,1].

Put simply, (NG) states that $F(q_1)$ has "no 'flat spots' of positive length" (Jeffrey, 1984, p. 82).

Using DFRT it can be proved that the predictive requirement (Reich) (see Chapter 2.1) is equivalent to (NG).[4]

5.2 CONVERGENCE OF OPINION AND CONVERGENCE TOWARDS THE TRUTH

DFRT *does* imply that a researcher *X* using an exchangeable inductive method *I behaves as if* (i) he believed that the process *Ex* under investigation were a (multivariate) Bernoulli process with parameter (vector) **q**, and (ii) his beliefs about the value of **q** were expressed by the *cdf* $F_I(q_1,\dots,q_{k-1})$ representing *I*. However, DFRT does *not* imply that *X should believe* that *Ex* is a multivariate Bernoulli process governed by a vector **q** of objective probabilities $\mathbf{q}_1,\dots,\mathbf{q}_k$.

Regarding this problem de Finetti sees $F_I(q_1,\dots,q_{k-1})$ as sheer 'mathematical fiction' relative to the 'fictitious entities' $\mathbf{q}_1,\dots,\mathbf{q}_k$. However this standpoint originates more from de Finetti's anti-realistic philosophical view that physical probabilities *do not* exist than from the mathematical content of DFRT.[5] Indeed, if the realist view that physical, or objective, probabilities *do* exist is accepted, there is no problem in assuming that *X may believe* that *Ex* is a multivariate Bernoulli process governed by a vector **q** of objective probabilities $\mathbf{q}_1,\dots,\mathbf{q}_k$.[6]

Different 'philosophical interpretations' of DFRT have different implications concerning the nature of 'scientific consensus', as will be seen below.

In almost all disciplines, the scientific community reveals a remarkable capacity to attain a considerable degree of consensus on the subjects under examination. Two questions need to be tackled: (1) which are the mechanisms governing the formation of this consensus? (2) to what extent is this consensus *rational*?

Many epistemologists and statisticians have dealt with the problems concerning the convergence of opinion among researchers who hold different initial beliefs.

For instance, in BS it has been pointed out that "a sufficiently large body of experience" can bring "holders of different opinions ... to similar opinions", i.e., to similar posterior distributions (Savage, 1972, pp. 68).

Regarding the issue of rational consensus among different scientists, de Finetti, as early as 1937 (p. 118), poses the question:

> Why are we obliged in the majority of problems to evaluate a probability according to the observation of a frequency?

Having remarked that this "is a question of the relations between the observation of past frequencies and the prediction of future frequencies" (*ibid.*), de Finetti advances several formulations (at least partially equivalent to each other) of "the requirement of induction":

> [the requirement of induction is] ... the requirement according to which the probability ought to be close to the observed frequency. (*Ibid.*, p. 120)

> ... a rich enough experience leads us always to consider as probable future frequencies ... close to those which have been observed. (*Ibid.*, p. 142)

> ... in the prediction of a frequency one is generally guided, or at least influenced, by the observation of past frequencies. (*Ibid.*, p. 152)

> ... the probability of a subsequent trial, relative to the observation of a certain frequency, tends to coincide with the value of the latter. (*Ibid.*, p. 153)

It is clear that the first and fourth formulations can be seen as informal anticipations of (Reich).

According to de Finetti (*ibid.*, p. 120) the requirement of induction is *not* a universally valid principle:

> It is evident that by posing the problem as we have, it will be impossible for us to *demonstrate* the validity of the requirement of induction ... That this requirement can only be justified in particular cases is not due to an insufficiency of the method followed ... Indeed, probability being purely subjective, nothing obliges us to choose it close to the frequency; all

that can be shown is that such an evaluation follows in a coherent manner from our initial judgment when the latter satisfies certain perfectly clear and natural conditions.

The above mentioned "perfectly clear and natural conditions" refer to the probability axioms and the condition of exchangeability. In fact, de Finetti (*ibid.*, pp. 142-147) argues that (normally) exchangeable probabilities do satisfy the requirement of induction, i.e., that exchangeable probabilities normally approach the observed relative frequencies. Given the convergence of exchangeable predictive probabilities towards the observed relative frequencies, the convergence of opinions among researchers with different initial beliefs follows directly. In this connection, de Finetti (*ibid.*, p. 147) points out that:

It is true that in many cases - as for example on the hypothesis of exchangeability - these subjective factors [i.e., different initial opinions] never have too pronounced an influence, provided that the experience be rich enough; the circumstance is very important, for it explains how in certain conditions more or less close agreement between the predictions of different individuals is produced.

According to de Finetti the process of learning from experience is no more than this convergence of predictive probabilities towards the observed relative frequencies. This assumption is strictly related to his anti-realistic view according to which physical probabilities do not exist and, consequently, there can be nothing to learn about them.

However if one accepts the realist view that physical, or objective, probabilities *do* exist, then a different account of learning from experience can be advanced. In fact, it could be argued that the process of learning does not consist *only* in the specific manner in which predictive probabilities are modified by experience; on the contrary, it *also* includes the tendency of predictive probabilities to *converge towards the true objective probabilities*. In other words, learning from experience is related to the fact that the predictive probabilities derived from a given inductive method tend to *approach the truth*. This signifies that a scientist using that particular inductive method tends to learn something about the real world.

Consider a (multivariate) Bernoulli process *Ex* with parameter (vector) **q**. Then a realist interpretation of "learning from experience" should include the claim that an inductive method I allows us to learn from experience iff, in any possible state of nature, i.e., for any value $q=(q_1,...,q_k)$ of **q**, the predictive probabilities $p(Q_i/e_n)$ specified by I tend to converge towards q_i in response to an increasingly large body of experience.

With reference to this, it should be pointed out that any exchangeable inductive method I which satisfies (Reich) makes learning from experience possible in the sense specified above. This can be proved as follows:

(a) if I satisfies (Reich), the predictive probabilities $p(Q_i/e_n)$ derived from I approach the empirical frequencies n_i/n (cf. Chapter 2.1);

(b) for any value q of $\mathbf{q}$, empirical frequencies n_i/n *converge almost surely* (i.e., with probability one) to the objective probabilities q_i, as stated by the strong law of large numbers of the probability calculus;

(c) hence, if I satisfies (Reich), then, for any value q of $\mathbf{q}$, the predictive probabilities $p(Q_i/e_n)$ derived from I converge almost surely to the objective probabilities q_i.

This signifies that an exchangeable inductive I method which satisfies (Reich) allows learning from experience not only in de Finetti's anti-realist sense but also in the realist sense that the predictive probabilities derived from I tend, *in the long run*, to approach the truth.[7] However, adopting (Reich) is of negligible help to a scientist who has to select an exchangeable inductive method since (Reich) is satisfied by 'almost all' exchangeable inductive methods.

It may be asked whether a scientist's concern with *approaching the truth in the short run* - i.e., on the basis of 'short' sequences of trials of *Ex* - can affect his choice of the inductive method. This question will be dealt with in Chapter 8.5 where the assumption that researchers attempt to maximize their possibilities of approaching the truth in the short run will be used to develop a 'verisimilitude approach' in selecting the optimum inductive method.

CHAPTER 6

GC-SYSTEMS AND DIRICHLET DISTRIBUTIONS

In this chapter a well known class of inductive methods, i.e., *GC-systems*, and a family of probability distributions widely used in BS, i.e., *Dirichlet distributions*, will be considered.

The main features of GC-systems and Dirichlet distributions are described in Sections 1-2. In Section 3 the relationships between the two are elucidated - more specifically it is shown that there is a one-to-one correspondence between the class of GC-systems and the class of the prior Dirichlet distributions on the parameter (vector) **q** of a (multivariate) Bernoulli process.

In Section 4 the 'extreme' GC-systems and the corresponding 'extreme' Dirichlet distributions are described.

Lastly, in Sections 5 and 6 some axiomatizations of GC-systems and Dirichlet distributions are considered.

6.1 GC-SYSTEMS

Generalized Carnapian systems, or *GC-systems*, are a class of inductive methods proposed by Carnap and Stegmüller (1959).[1] Such methods are defined as follows:

(1) A GC-system is an inductive method characterized by the special values:

$$p(Q_i/e_n) = \frac{(n_i + \gamma_i \lambda)}{(n + \lambda)} \quad \text{where } \gamma_i > 0, \Sigma\gamma_i = 1, \text{and } 0 < \lambda < \infty.$$

An interpretation of the parameter λ is given after formula (4) below (see also Chapter 7, note 18).

The vector $\gamma=(\gamma_1,...,\gamma_k)$ related to a given GC-system will be called the *prior vector* of that system since it follows from (1), by putting $n_i = n = 0$, that GC-systems satisfy the following *requirement of initial possibility*:

(IP) $p(Q_i) = \gamma_i > 0$

Since a given GC-system is fully characterized by a couple (γ,λ) it may be referred to as (γ,λ). To emphasize that a special value $p(Q_i/e_n)$ belongs to a specific GC-system (γ,λ) it may be denoted by $p_{\gamma,\lambda}(Q_i/e_n)$.

The inductive methods introduced by Carnap (1952), which will be referred to as *Carnapian systems*, or *C-systems*, can be defined as follows:[2]

(2) A C-system is an inductive method characterized by the special values:

$$p(Q_i/e_n) = \frac{(n_i + \lambda/k)}{(n + \lambda)} \qquad \text{where } 0 < \lambda < \infty.$$

A C-system can be seen as a GC-system with prior vector γ equal to $\mathbf{1/k}=(1/k,...,1/k)$. In other words, a C-system is a GC-system which satisfies, besides (IP), the stronger *requirement of initial equipossibility* (IE):

(IE) $p(Q_i) = 1/k$

GC-systems satisfy all the basic requirements of TIP as illustrated in Chapter 2:

(3) GC-systems satisfy (Exc), (Reg), (PR) and (Reich).

From (1) it can be seen that the special values $p(Q_i/e_n)$ of a GC-system depend on the empirical evidence e_n only via n and n_i. More specifically, GC-systems satisfy the following *requirement of restricted relevance (RR)*:

(RR) If e_n and e'_n are such that $n_i = n'_i$, then $p(Q_i/e_n) = p(Q_i/e'_n)$.

The requirement (RR) implies that the predictive probabilities $p(Q_i/e_n)$ should in no way be affected by the empirical frequencies n_j of Q_j in the sequence e_n - where $Q_j \neq Q_i$ - independently of the possible, more or less strong, similarity occurring between the properties Q_j and Q_i. This signifies that 'analogy by similarity' is outside the scope of GC-systems.[3]

The *posterior vector* $(p(Q_1/e_n),...,p(Q_k/e_n))$ of a GC-system is obtained by a simple combination of the prior information contained in the *prior vector* $\gamma=(\gamma_1,...,\gamma_k)$ and the empirical information contained in the *empirical vector* $(n_1/n,...,n_k/n)$. Indeed, the predictive probabilities $p(Q_i/e_n)$ of a GC-system are a *weighted average* of the corresponding prior probabilities γ_i and empirical frequencies n_i/n, where $\lambda/(n+\lambda)$ is the *weight of the prior probability* γ_i and $n/(n+\lambda)$ is the *weight of the empirical frequency* n_i/n (cf. Carnap, 1980, p. 94):

$$(4) \qquad p(Q_i/e_n) = \frac{\lambda}{n+\lambda}\gamma_i + \frac{n}{n+\lambda}\cdot\frac{n_i}{n}$$

It can be seen that the weight $n/(n+\lambda)$ of the empirical frequency n_i/n increases, as one would naturally expect, with the cardinality n of the empirical sample: this weight is 0 when $n = 0$ and approaches its maximum value 1 when n becomes much higher than λ.

GC-systems can be seen as a natural modification of the so-called *straight rule* $p_0(Q_i/e_n) = n_i/n$ which equates the posterior predictive probability straight away to the relative frequency n_i/n of Q_i in e_n (cf. Carnap, 1980, p. 85). In fact the special values $p(Q_i/e_n)$ occurring in (1) may be obtained from $p_0(Q_i/e_n)$ by augmenting "the n *real* individuals with λ *virtual* individuals, among which $\lambda\gamma_i$ have the property Q_i for each $i = 1,...,k$. Then the value of $p(Q_i/e_n)$ ... is simply the relative frequency of Q_i among the $n + \lambda$ *real and virtual* individuals" (Jeffrey, 1980, pp. 2-3).

Let $p(Q_i/Q_j)$ be the predictive probability of Q_i relative to the empirical evidence given by the observation of a single Q_j-outcome. Then, for any couple of distinct attribute indices i, j, the *relevance quotient* η_{ij} is defined as follows:

$$(5) \qquad \eta_{ij} = \frac{p(Q_i/Q_j)}{p(Q_i)}$$

The relevance quotient η_{ij} represents the proportion of the prior probability $p(Q_i)$ of Q_i which 'survives' after the observation of the first 'negative example' Q_j. An important property of GC-systems is the so-called η-*equality* (Carnap, 1980, p. 57):

$$(6) \qquad \eta_{ij} = \eta = \frac{\lambda}{(1+\lambda)} \qquad \text{where } 0 < \eta < 1.$$

6.2 DIRICHLET DISTRIBUTIONS

Beta distributions

A *Beta distribution* Beta(α_1,α_2) on a parameter $\mathbf{q}_1$, where $0 \leq \mathbf{q}_1 \leq 1$, is characterized by a *pdf* Beta(q_1;α_1,α_2) defined as follows:

(7) $$\mathrm{Beta}(q_1;\alpha_1,\alpha_2) = \frac{\Gamma(\alpha_1 + \alpha_2)}{\Gamma(\alpha_1)\,\Gamma(\alpha_2)}\, q_1^{\alpha_1-1}(1-q_1)^{\alpha_2-1} \quad \text{where } \alpha_1, \alpha_2 > 0.$$ [4]

$\Gamma(\alpha_1+\alpha_2)/[\Gamma(\alpha_1)\Gamma(\alpha_2)]$ is simply a 'normalizing factor' employed to render the total probability on $\mathbf{q}_1$ equal to 1. (Here $\Gamma(x)$ denotes the so-called *gamma function* - or *generalized factorial* function - which is defined for all positive real numbers x and has the value of an ordinary factorial for any integer number n: $\Gamma(n) = (n - 1)!$.)

The mean and the variance of Beta(α_1,α_2) are given by:

(8) $$E(\mathbf{q}_1) = \frac{\alpha_1}{\alpha_1 + \alpha_2}$$

(9) $$\mathrm{var}(\mathbf{q}_1) = \frac{\alpha_1\alpha_2}{(\alpha_1 + \alpha_2)^2(\alpha_1 + \alpha_2 + 1)}$$

Beta priors are *conjugate priors* for the parameter $\mathbf{q}_1$ of a Bernoulli process. Essentially this means that if the prior distribution on $\mathbf{q}_1$ is a Beta then the posterior distribution, relative to e_n, is also a Beta. More specifically:

(10) If $f(q_1) = \mathrm{Beta}(q_1;\alpha_1,\alpha_2)$, then $f(q_1/e_n) = \mathrm{Beta}(q_1;\alpha_1+n_1,\alpha_2+n-n_1)$.

From (8) and (10) it follows that, given the prior Beta(α_1,α_2), the posterior expectation $E(\mathbf{q}_1 \mid e_n)$ is given by:

(11) $$E(\mathbf{q}_1 \mid e_n) = \frac{\alpha_1 + n_1}{\alpha_1 + \alpha_2 + n}$$

A Beta distribution Beta(α_1,α_2) with parameters $\alpha_1 = \alpha_2 = \alpha^*$ is called a *symmetrical Beta distribution* since its mean $E(\mathbf{q}_1)$ is 1/2 (cf. (8)). In particular: (i) when $\alpha^* < 1$, a U-shaped beta distribution Beta(α^*,α^*) with very high variance is obtained; (ii) when $\alpha^* = 1$, the beta distribution Beta(1,1) identical to the so-called *uniform distribution* on $\mathbf{q}_1$ is obtained;[5]

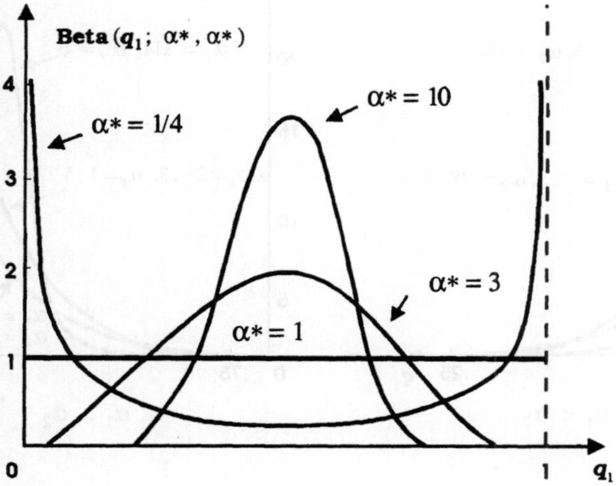

Figure 1

Density functions of symmetrical Beta distributions

(iii) when $\alpha^* >> 1$, a beta distribution Beta(α^*,α^*) with small variance is obtained (see Figure 1: cf. Winkler, 1972, p. 151).

A Beta distribution Beta(α_1,α_2) with parameters $\alpha_1 \neq \alpha_2$ is called an *asymmetrical Beta distribution*. Some of the shapes which may be assumed by the probability densities of asymmetrical Beta distributions are shown in Figure 2 (cf. Winkler, *ibid.*, pp. 151-152)

The above observations demonstrate that the Beta family is very "rich" in the sense that it includes "distributions with different locations, dispersions, shapes .... so as to be able to represent a wide variety of states of prior information" (Winkler, *ibid.*, p. 148).

Dirichlet distributions

The multivariate analogues of Beta distributions for parameter vectors $\mathbf{q}=(\mathbf{q}_1,...,\mathbf{q}_k)$ are the so-called *Dirichlet distributions* Dir($\alpha_1,...,\alpha_k$). The *pdf* Dir($q_1,...,q_{k-1};\alpha_1,...,\alpha_k$) of a k-1-variate Dirichlet distribution is defined as

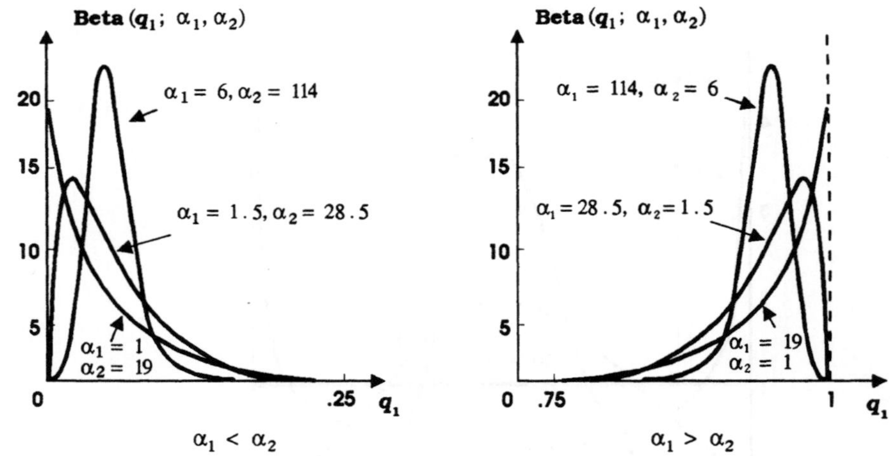

Figure 2

Probability densities of asymmetrical Beta distributions

follows:

(12) $$\mathrm{Dir}(q_1,\ldots,q_{k-1};\alpha_1,\ldots,\alpha_k) = \frac{\Gamma(\alpha)}{\Gamma(\alpha_1)\cdots\Gamma(\alpha_k)}\, q_1^{\alpha_1-1}\cdots q_k^{\alpha_k-1}$$

where $\alpha_1,\ldots,\alpha_k > 0$, $\alpha = \Sigma\alpha_i$, and $q_k = 1 - q_1 - \cdots - q_{k-1}$.[6]

The constant $\Gamma(\alpha)/[\Gamma(\alpha_1)\cdots\Gamma(\alpha_k)]$ is again a normalizing factor rendering the total probability equal to 1.

Notice that the marginal distributions of a Dirichlet distribution are Beta distributions:

(13) If $f(q_1,\ldots,q_{k-1}) = \mathrm{Dir}(q_1,\ldots,q_{k-1};\alpha_1,\ldots,\alpha_k)$ then $f(q_i) = \mathrm{Beta}(q_i;\beta_1,\beta_2)$ where $\beta_1 = \alpha_i$ and $\beta_2 = \alpha - \alpha_i$.

The marginal means of $\mathrm{Dir}(\alpha_1,\ldots,\alpha_k)$ are given by:

(14) $E(\mathbf{q}_i) = \alpha_i/\alpha = \gamma_i$

Other typical values of Dirichlet distributions are given by the following formulae:

(15) $$E(\mathbf{q}_i^2) = \frac{\alpha_i(\alpha - \alpha_i)}{\alpha(\alpha + 1)}$$

(16) $$\text{var}(\mathbf{q}_i) = \frac{\alpha_i(\alpha - \alpha_i)}{\alpha^2(\alpha + 1)}$$

(17) $$E(\mathbf{q}_i\mathbf{q}_j) = \frac{\alpha_i\alpha_j}{\alpha(\alpha + 1)}$$

(18) $$\text{cov}(\mathbf{q}_i, \mathbf{q}_j) = -\frac{\alpha_i\alpha_j}{\alpha^2(\alpha + 1)}$$

As a consequence of (16), the value of $\text{VAR}(\mathbf{q}) = \Sigma\text{var}(\mathbf{q}_i)$ (cf. (3.37)) is given by:

(19) $$\text{VAR}(\mathbf{q}) = \frac{\alpha^2 - \Sigma\alpha_i^2}{\alpha^2(\alpha + 1)}$$

It follows from the general property (3.32) of the *cdf*'s $F(q_1,\ldots,q_{k-1})$ that the prior predictive probability $p(Q_i)$ derived from a Dirichlet distribution $\text{Dir}(\alpha_1,\ldots,\alpha_k)$ is identical to its marginal mean $E(\mathbf{q}_i)$. Here, $E(\mathbf{q}_i)$ (= $p(Q_i)$) will be denoted by the shorter γ_i and the mean vector $E(\mathbf{q})=(E(\mathbf{q}_1),\ldots, E(\mathbf{q}_k))$ by $\gamma=(\gamma_1,\ldots,\gamma_k)$. These symbols have already been used to denote the prior probabilities of GC-systems (see Section 1). The vector $\gamma=(\gamma_1,\ldots,\gamma_k)$ will hence be referred to as the *prior vector* of $\text{Dir}(\alpha_1,\ldots,\alpha_k)$.

Given that $\alpha_i = \alpha E(\mathbf{q}_i) = \alpha\gamma_i$ (cf. (14)), formulae (15), (16), (18) and (19) can be rewritten as follows:

(20) $$E(\mathbf{q}_i^2) = \frac{\gamma_i(\alpha\gamma_i + 1)}{\alpha + 1}$$

(21) $$\mathrm{var}(\mathbf{q_i}) = \frac{\gamma_i(1-\gamma_i)}{\alpha+1}$$

(22) $$\mathrm{cov}(\mathbf{q_i},\mathbf{q_j}) = -\frac{\gamma_i\gamma_j}{\alpha+1}$$

(23) $$\mathrm{VAR}(\mathbf{q}) = \frac{1-\Sigma\gamma_i^2}{\alpha+1} = \frac{G(\gamma)}{\alpha+1}$$

The value of $E[G(\mathbf{q})] = 1 - \Sigma E(\mathbf{q_i^2})$ is derived from (20):

(24) $$E[G(\mathbf{q})] = \frac{\alpha}{\alpha+1}G(\gamma)$$

where $G(\gamma)$ $(= 1 - \Sigma\gamma_i^2)$ is the diversity of the prior vector γ.

Remember that the expected diversity $E[G(\mathbf{q})]$ of any *cdf* $F(q_1,\dots,q_{k-1})$ is included in the interval $[0,G(\gamma)]$ (see (3.43)). In particular, regarding Dirichlet distributions $\mathrm{Dir}(\alpha_1,\dots,\alpha_k)$, it follows from (24) and from the inequalities $0 < \alpha < \infty$, that:

(25) $E[G(\mathbf{q})]$ is included in the interval $[0,G(\gamma)]$.
More specifically: when $\alpha \to 0$ then $E[G(\mathbf{q})] \to 0$ and when $\alpha \to \infty$ then $E[G(\mathbf{q})] \to G(\gamma)$.

Analogous to the Beta distributions, the Dirichlet distributions are conjugate distributions for the parameter vector $\mathbf{q}=(\mathbf{q_1},\dots,\mathbf{q_k})$ of a multivariate Bernoulli process. More specifically:

(26) If $f(q_1,\dots,q_{k-1}) = \mathrm{Dir}(q_1,\dots,q_{k-1};\alpha_1,\dots,\alpha_k)$,
then $f(q_1,\dots,q_{k-1}/e_n) = \mathrm{Dir}(q_1,\dots,q_{k-1};\alpha_1+n_1,\dots,\alpha_k+n_k)$.

From (26), (13) and (11) it follows that the posterior marginal mean $E(\mathbf{q_i} \mid e_n)$ of $\mathrm{Dir}(\alpha_1,\dots,\alpha_k)$ is given by:

(27) $$E(\mathbf{q_i} \mid e_n) = \frac{\alpha_i+n_i}{\alpha+n}$$

A Dirichlet distribution $\mathrm{Dir}(\alpha_1,\ldots,\alpha_k)$ with parameters $\alpha_1 = \cdots = \alpha_k = \alpha^*$, where $\alpha = k\alpha^*$, is called a *symmetrical Dirichlet distribution* since the marginal means $E(\mathbf{q}_i)$ are equal to $1/k$ (see (14)). The *pdf* of a symmetrical Dirichlet distribution $\mathrm{Dir}(\alpha^*,\ldots,\alpha^*)$ has the following form (cf. Good, 1965, p. 25):

(28) $$\mathrm{Dir}(q_1,\ldots,q_{k-1};\alpha^*,\ldots,\alpha^*) = \frac{\Gamma(k\alpha^*)}{[\Gamma(\alpha^*)]^k}\, q_1^{\alpha^*-1} \cdots q_k^{\alpha^*-1}$$

In particular, when $\alpha^* = 1$, the Dirichlet distribution $\mathrm{Dir}(1,\ldots,1)$ is identical to the so-called *uniform distribution* on the parameter vector **q**.[7]

6.3 THE EQUIVALENCE BETWEEN GC-SYSTEMS AND DIRICHLET DISTRIBUTIONS

Given that $E(\mathbf{q}_i \mid e_n) = p(Q_i/e_n)$ (see (3.32)) and $\alpha_i = \alpha\gamma_i$ (see (14)), it follows from (27) that the special values $p(Q_i/e_n)$ derived from $\mathrm{Dir}(\alpha_1,\ldots,\alpha_k)$ are given by:

(29) $$p(Q_i/e_n) = \frac{\alpha\gamma_i + n_i}{\alpha + n}$$

It follows from (29) and from the definition (1) of GC-systems that:

(30) Given the prior $\mathrm{Dir}(\alpha_1,\ldots,\alpha_k)$, the derived set of predictive probabilities $p(Q_i/e_n)$ is a GC-system (γ,α) where $\gamma=(\gamma_1,\ldots,\gamma_k) = (\alpha_1/\alpha,\ldots,\alpha_k/\alpha)$.

As a direct consequence of (30):

(31) The GC-system (γ,λ) can be derived from the corresponding prior $\mathrm{Dir}(\gamma_1\lambda,\ldots,\gamma_k\lambda)$.

Given that GC-systems are exchangeable inductive methods, it follows from (31) and from the uniqueness clause of DFRT (see (5.2)), that:

(32) The GC-system (γ,λ) can be derived from a *unique* prior distribution on the parameter vector **q** where this prior is $\mathrm{Dir}(\gamma_1\lambda,\ldots,\gamma_k\lambda)$.

It follows from (30) and (32) that:

(33) There is a one-to-one correspondence between the class of GC-systems (γ,λ) and the class of the prior Dirichlet $\mathrm{Dir}(\alpha_1,\ldots,\alpha_k)$ on the parameter vector **q** of a multivariate Bernoulli process. The parameters are related as follows: $\lambda = \alpha = \Sigma\alpha_i$ and $\gamma_i = \alpha_i/\alpha$.

Two corollaries of (33) are: (i) there is a one-to-one correspondence between the class of C-systems $(\mathbf{1/k},\lambda)$ and the class of the symmetrical Dirichlet distributions $\mathrm{Dir}(\lambda/k,\ldots,\lambda/k)$; (ii) there is a one-to-one correspondence between the class of GC-systems (γ,λ) with $\gamma=(\gamma_1,\gamma_2)$ and the class of the Beta distributions $\mathrm{Beta}(\gamma_1\lambda,\gamma_2\lambda)$ on the parameter $\mathbf{q}_1$ of a Bernoulli process.[8]

6.4 EXTREME GC-SYSTEMS AND EXTREME DIRICHLET DISTRIBUTIONS

Consider the set of GC-systems (γ,λ) with a given prior vector γ. Then the inductive methods $(\gamma,0)$ and (γ,∞) are characterized by the special values $p_{\gamma,0}(Q_i/e_n)$ and $p_{\gamma,\infty}(Q_i/e_n)$ which are defined thus (cf. Carnap, 1980, p. 95):

(34) $$p_{\gamma,0}(Q_i/e_n) = \lim_{\lambda \to 0} p_{\gamma,\lambda}(Q_i/e_n)$$

(35) $$p_{\gamma,\infty}(Q_i/e_n) = \lim_{\lambda \to \infty} p_{\gamma,\lambda}(Q_i/e_n)$$

It follows from (34) and (1) that $(\gamma,0)$ is almost identical to the above mentioned straight rule p_0 since, for any $n \neq 0$, $p_{\gamma,0}(Q_i/e_n) = n_i/n$.[9] Since the special values $p_{\gamma,0}(Q_i/e_n)$ are equated to the corresponding empirical frequencies and no weight is given to prior probabilities, $(\gamma,0)$ can be seen as an *extreme* inductive method characterized by an extreme propensity to learn from experience.[10]

As far as (γ,∞) is concerned, it follows from (35) and (1) that $p_{\gamma,\infty}(Q_i/e_n) = \gamma_i$. Since the special values $p_{\gamma,\infty}(Q_i/e_n)$ are equal to the corresponding prior probabilities γ_i and no weight is given to the empirical frequencies, (γ,∞) may be seen as a *non-inductive* method since whoever adopts (γ,∞) "refuses to learn anything from experience" (Carnap, *ibid.*, p. 95).

Although, strictly speaking, $(\gamma,0)$ and (γ,∞) are not GC-systems (cf. (1)), they may be considered as *extreme* GC-systems (γ,λ), where $\lambda = 0$ and $\lambda = \infty$.

It can be shown that $(\gamma,0)$ and (γ,∞) are, in a certain sense, equivalent to two types of prior distributions on the parameter vector $\mathbf{q}$ which could be called *extreme Dirichlet distributions.*

For this purpose it should be recalled that an *improper distribution* is one which assigns infinite 'probability' to the whole range of a given parameter (vector) but finite probability to any compact (i.e., bounded and closed) subset of it. It has been argued that improper distributions may be used as priors in Bayesian inferences given that the corresponding posterior distributions derived using Bayes's theorem are proper (cf. Novick and Hall, 1965).

Consider, for instance, the distribution characterized by the density:

(36) $$\prod_{i=1}^{k} q_i^{-1}$$

It can be proved that this distribution is improper since the integral of $\Pi\, q_i^{-1}$ on the range of $\mathbf{q}$ diverges (cf. Good, 1965, p. 28).

Since $\Pi\, q_i^{-1}$ has the algebraic form of the density of a Dirichlet distribution (see (12)) - except for the values $\alpha_i = 0$ of the parameters - it may be seen as an *improper Dirichlet distribution* and referred to as Dir(0,...,0). It can be proved that:

(37) If $f(q_1,...,q_{k-1}) = \mathrm{Dir}(q_1,...,q_{k-1};0,...,0)$,
then $f(q_1,...,q_{k-1}/e_n) = \mathrm{Dir}(q_1,...,q_{k-1};n_1,...,n_k)$,

where $\mathrm{Dir}(q_1,...,q_{k-1};n_1,...,n_k)$ is proper, provided that all n_i $(i = 1,...,k)$ are positive.

From (37), (13) and (14) it follows that the posterior marginal means $E(\mathbf{q}_i \mid e_n)$ $(= p(Q_i/e_n))$ of Dir(0,...,0) are equal to n_i/n. Therefore Dir(0,...,0) is virtually equivalent to the extreme GC-system $(\gamma,0)$.

Lastly, consider a 'degenerate' prior $F(q_1,...,q_{k-1})$ attributing the whole probability to a single value $(\gamma_1,...,\gamma_{k-1})$ of the parameter vector $\mathbf{q}$. In this case the prior and posterior marginal means and, accordingly, the prior and posterior predictive probabilities $p(Q_i/e_n)$ derived from $F(q_1,...,q_{k-1})$ are identical to γ_i. Hence, this distribution is equivalent to the extreme GC-system (γ,∞). The prior *cdf* $F(q_1,...,q_{k-1})$ may be called a *degenerate Dirichlet distribution*, since it can be seen as the limit of a class of Dirichlet priors $\mathrm{Dir}(\alpha_1,...,\alpha_k)$ whose parameters α_i tend to ∞, while the ratios $\gamma_i = \alpha_i/\alpha$ remain unchanged. In fact, when $\alpha_1,...,\alpha_k$ and, hence, α tend to ∞, the marginal variances of $\mathrm{Dir}(\alpha_1,...,\alpha_k)$ tend to zero (cf. (21)) and the distribution concentrates almost entirely on the prior vector $(\gamma_1,...,\gamma_{k-1})$. Consequently $\mathrm{Dir}(\alpha_1,...,\alpha_k)$ becomes virtually indistinguishable from the particular $F(q_1,...,q_{k-1})$ we started with.

6.5 THE AXIOMATIZATION OF GC-SYSTEMS

Axiomatizing a given class of inductive methods is equal to proving that the class includes *all* and *only* those methods which satisfy a given set of axioms. In particular, C- and GC-systems have been axiomatized by several authors, e.g. Carnap and Stegmüller (1959), Kemeny (1964), Carnap (1980, § 19), and Kuipers (1978, Ch. 5). Their results may be summarized as follows:

(38) An inductive method for a family $(Q_1,...,Q_k)$, where $k > 2$, is a GC-system iff it satisfies the requirements of exchangeability (Exc), of restricted relevance (RR), and of initial possibility (IP).[11]

They have also proved that, when $k > 2$, an inductive method is a C-system iff it satisfies (Exc), (RR) and the requirement of initial equipossibility (IE).[12]

Whilst it is very simple to prove that all GC-systems satisfy (Exc), (RR) and (IP) (see Section 1), proving that these requirements are satisfied only by GC-systems is anything but easy. This proof is essentially based on the following lemma:

(39) Given the axioms of probability calculus and (Exc), for $k > 2$ the requirement of restricted relevance (RR) is equivalent to the following *requirement of linearity*:

(RL) $p(Q_i/e_n)$ is a linear function of n_i. More precisely, there are constants $a_i > 0$ and b such that $p(Q_i/e_n) = a_i + bn_i$.

Unfortunately, this strategy cannot be applied when $k = 2$ since in this case (RR) and (RL) *are not* equivalent.[13] For this reason, where $k = 2$, the pre-mentioned authors directly include (RL) among the axioms:

(40) An inductive method for a family (Q_1,Q_2) is a GC-system iff it satisfies (Exc), (RL), and (IP).

They have also proved that, where $k = 2$, an inductive method is a C-system iff it satisfies (Exc), (RL) and (IE).

In place of this 'double' axiomatization, "it would be desirable" - Carnap (1980, p. 101) remarks - "to have a single assumption applicable to any $k \geq 2$, one that would yield the same results as our present two assumptions together".[14]

As far as I know, the only available 'unified axiomatization' of GC-systems was proposed by Costantini (1979) and employs a "single assumption" called the *condition of invariance of the relevance quotient*, or *(CIRQ)*.

Given the sequence of trials e_n, and the sequence $e_n Q_i$ which is obtained by adding a Q_i-outcome to e_n, Costantini (*ibid.*, p. 151) defines the *relevance quotient* $R_{ij}(e_n)$ *for* Q_i *and* Q_j *w.r.t.* e_n thus:

(41) $$R_{ij}(e_n) = \frac{p(Q_i/e_n Q_j)}{p(Q_i/e_n)}$$

The relevance quotient $R_{ij}(e_n)$ represents the proportion of the predictive probability $p(Q_i/e_n)$ which 'survives' after the observation of a 'negative example' Q_j. The relevance quotient can be seen as a generalization of Carnap's relevance quotient η_{ij} since it follows from (5) that $\eta_{ij} = p(Q_i/Q_j)/p(Q_i) = R_{ij}(e_0)$, where e_0 denotes an 'empty sequence', or tautological evidence.

Now (CIRQ) is a natural generalization of the η-equality $\eta_{ij} = \eta$ (cf. (6)):

(CIRQ) $R_{ij}(e_n)$ - where $i \neq j$ - depends only on n.

(CIRQ) demands that $R_{ij}(e_n)$ should depend neither on the particular sequence e_n nor on the particular couple Q_i, Q_j (where $i \neq j$) but exclusively on the 'length' n of e_n.[15]

The axiomatization of GC-systems proposed by Costantini (*ibid.*, pp. 154-156) may be summarized as follows:[16]

(42) An inductive method for a family $(Q_1,...,Q_k)$, where $k \geq 2$, is a GC-system iff it satisfies (Exc), (CIRQ), and (IP).

6.6 THE AXIOMATIZATION OF DIRICHLET DISTRIBUTIONS

Axiomatizing a given class of probability distributions for a given parameter (vector) is equal to proving that the class includes *all* and *only* those distributions which satisfy a given set of parametric requirements used as axioms (cf. Chapter 5.1).

In particular, certain parametric requirements which use the concept of *neutrality*, as introduced by Connor and Mosimann (1969, pp. 194-199), are applied to axiomatize Dirichlet distributions.

Put simply, given the *proportions* $\mathbf{q}_1,...,\mathbf{q}_k$ - which are defined as non-negative parameters satisfying the constraint "$\Sigma\mathbf{q}_i = 1$" - the proportion $\mathbf{q}_1$ is said to be *neutral* w.r.t. $\mathbf{q}=(\mathbf{q}_1,...,\mathbf{q}_k)$ when no information on the value of $\mathbf{q}_1$ can influence our beliefs concerning "the manner in which the remaining proportions $\mathbf{q}_2,...,\mathbf{q}_k$ *proportionally* divide the remainder of the unit interval" (Connor and Mosimann, *ibid.*, p. 196).[17]

A precise formulation of the concept of neutrality is given by Fabius (1973, p. 583) who defines the notion of *(CM)$_i$-neutral vector* as follows:

(43) Given a *cdf* $F(q_1,...,q_{k-1})$, the vector $\mathbf{q}=(\mathbf{q}_1,...,\mathbf{q}_k)$ is *(CM)$_i$-neutral* iff the fractions $\mathbf{q}_j/(1 - \mathbf{q}_i)$, where $j \neq i$, are independent of $\mathbf{q}_i$.

Recall that $\mathbf{q}_j/(1 - \mathbf{q}_i)$ is independent of $\mathbf{q}_i$, w.r.t. a *cdf* $F(q_1,...,q_{k-1})$ on $\mathbf{q}$, iff $F(q_j/(1 - q_i),q_i) = F(q_j/(1 - q_i))F(q_i)$ (cf. definition (3.4)).

The following (parametric) requirement of neutrality (CM) on a *cdf* $F(q_1,...,q_{k-1})$ can be stated by using the concept of (CM)$_i$-neutrality:

(CM) Given a *cdf* $F(q_1,...,q_{k-1})$, the vector $\mathbf{q}=(\mathbf{q}_1,...,\mathbf{q}_k)$ is *(CM)$_i$*-neutral for $i = 1,...,k$.

The intuitive content of (CM) is akin to that of the predictive requirement (RR) used to axiomatize GC-systems. In fact, as seen above, (RR) excludes any 'analogy by similarity' since it requires that the predictive probabilities $p(Q_i/e_n)$ should in no way be affected by the empirical frequencies n_j of Q_j in e_n - where $Q_j \neq Q_i$ - independently of the possible, more or less strong, similarity occurring between the properties Q_j and Q_i. Now, in a certain sense, (CM) also excludes any 'analogy by similarity' since it requires that our beliefs about $\mathbf{q}_i$ should in no way be affected by any information concerning the value of the fraction $\mathbf{q}_j/(1 - \mathbf{q}_i)$ in the remainder $(1 - \mathbf{q}_i)$ of the unit interval; this independently of the possible, more or less strong, similarity occurring between the properties Q_j and Q_i corresponding to $\mathbf{q}_i$ and $\mathbf{q}_j$.

The parametric requirement (CM) forms the basis of Fabius's axiomatization (1973) of Dirichlet distributions (*ibid.*) which can be stated as follows:[18]

(44) A *cdf* $F(q_1,...,q_{k-1})$ on $\mathbf{q}=(\mathbf{q}_1,...,\mathbf{q}_k)$ is a Dirichlet distribution iff it satisfies (CM).

The formulation of Fabius's theorem given in (44) is somewhat imprecise. Fabius (*ibid.*, p. 583), indeed, proves that a probability distribution on **q** satisfies (CM) iff "*the distribution ... is a Dirichlet distribution or a limit of Dirichlet distribution*" which "is just a short way of saying that the distribution ... is either a Dirichlet distribution, or discrete and concentrated in the vertices of the simplex in which [**q**] takes its values, or degenerate" (*ibid.*, p. 584).

This axiomatization of Dirichlet distributions does not apply when $k = 2$, where the parametric principle (CM) is trivially true. As far as I know a unified axiomatic characterization of Beta and Dirichlet distribution on the basis of appropriate parametric requirements has still to be found.

PART III

VERISIMILITUDE, DISORDER, AND OPTIMUM PRIOR PROBABILITIES

CHAPTER 7

THE CHOICE OF PRIOR PROBABILITIES: THE SUBJECTIVE, APRIORISTIC, AND CONTEXTUAL APPROACHES

In Bayesian methodology the posterior probabilities which a scientist attributes to the hypotheses under investigation depend on his prior information (prior beliefs) and the new empirical information obtained from a given experiment (cf. Chapter 3.1). Given a sufficiently large body of experimental data, different scientists with different prior beliefs are led to attribute similar posterior probabilities (cf. Chapter 5.2). Conversely, when the available experimental evidence is limited, then the posterior probabilities of different scientists will depend to a large extent on their prior beliefs, i.e., on their prior probabilities. In such cases, the degree of consensus about posterior probabilities arrived at by the scientific community depends almost entirely on the corresponding degree of consensus in selecting prior probabilities.

In this chapter three competing approaches to the problem of the choice of priors probabilities are considered.

In Sections 1-4 the two 'traditional' approaches to this problem, i.e., the *subjective approach* and the *aprioristic approach*, are considered. In Section 1 and 2 these approaches are considered w.r.t. the choice of priors in BS. In Sections 3 and 4 the subjective and the aprioristic interpretations of TIP are examined.

In Section 5 a 'new' *contextual approach* to the selection of prior probabilities is presented.

Finally, in Section 6 a context-dependent justification of the decision to use a prior Dirichlet distribution or, equivalently, a GC-system, within a given inquiry is proposed.

7.1 THE CHOICE OF PRIORS IN BAYESIAN STATISTICS: THE SUBJECTIVE APPROACH

According to the *subjective approach* to Bayesian inference the probability axioms are the only principles of rationality governing the selection of prior probabilities in Bayesian inferences. Subjective Bayesians recognize that some '*descriptive principles*' capture important features of the beliefs of (almost) all researchers engaged in certain types of empirical researches. For instance, Savage (1972, pp. 63-67) points out the role of symmetry in a number of situations where the probability judgments of different individuals typically tend to agree. However, subjective Bayesians reject the idea that epistemic probabilities are ruled by other principles of rationality - or *normative principles* - besides the probability axioms.

A good example of this view is given by de Finetti's interpretation of (Exc). De Finetti does not consider (Exc) as a normative principle subject to rational justification. Although he claims that his "hypothesis of exchangeability" offers a convincing "justification" - or "explanation" - of the "principle of induction" (cf. Chapter 5.2), he is very clear about the "purely psychological" character of this justification (de Finetti, 1937, p. 152). Indeed, de Finetti's (*ibid.*) objective is to show that:

> *... there are rather profound psychological reasons which make the exact or approximate agreement that is observed between the opinions of different individuals very natural, but that there are no reasons, rational, positive, or metaphysical, that can give this fact any meaning beyond that of a simple agreement of subjective opinions.*

Since the only principles of rationality admitted by subjective Bayesians are the probability axioms, in their view there is no place for the problem of the 'rational choice' of a prior distribution. As Winkler (1967, p. 777) remarks:

> The personalistic view [i.e., the subjective approach] differs from other approaches by not attempting to specify what assessments are 'correct'. All self-consistent, or coherent assessments are admissible as long as the individual feels that they correspond with his judgments.

That means that in subjective Bayesianism the problem of the choice of the priors is identical to the problem of selecting the prior distribution best fitting the prior beliefs of the researcher.

Several techniques have been developed to "bring to light" the actual beliefs of a given person by using a suitable "verbal" or "behavioral interrogation" (Savage, 1972, pp. 27-29). With reference to this, Winkler (*ibid.*, p. 778) says that:

What is needed is ... an interviewing procedure, with which one can "interview" the assessor and elicit enough information to be able to write down a prior distribution which accurately reflects the assessor's prior knowledge and judgments. It must be stressed that the assessor has no built-in prior distribution which is there for the taking. Rather, the assessor has certain prior knowledge which is not easy to express quantitatively without careful thought. An elicitation technique used by the statistician does not elicit a "true" prior distribution, but in a sense helps to draw out an assessment of a prior distribution from the prior knowledge. Different techniques may produce different distributions because the method of questioning may have some effect on the way the problem is viewed.

An efficient "interviewing procedure" should assess the prior distribution of a given subject with the minimal amount of (self-)interrogation.[1]

Suppose that the Beta prior Beta(α_1,α_2) best fitting a researcher *X*'s prior beliefs about the parameter $\mathbf{q}_1$ of a Bernoulli process *Ex* has to be assessed. For this purpose the values of the mean and the variance of $\mathbf{q}_1$ could be assessed and, subsequently, the unique Beta with these two values could be determined using formulae (6.8) and (6.9).[2]

A similar strategy could be used where the Dirichlet prior Dir(α_1,...,α_k) best fitting *X*'s prior beliefs about the parameter vector $\mathbf{q} = (\mathbf{q}_1,...,\mathbf{q}_k)$ of a multivariate Bernoulli process *Ex* has to be assessed. We could first try to specify the prior vector $\gamma=(\gamma_1,...,\gamma_k)$ by assessing *X*'s expectations $E(\mathbf{q}_i) = \gamma_i$; secondly, we could assess *X*'s expectation $E[G(\mathbf{q})]$ of Gini diversity $G(\mathbf{q})$; thirdly, the value of α can be found using the following formula derived from (6.24):

$$(1) \qquad \alpha = \frac{E[G(\mathbf{q})]}{G(\gamma) - E[G(\mathbf{q})]}$$

where $G(\gamma)$ is the Gini diversity of the prior vector γ.

Lastly, the values of $\alpha_1,...,\alpha_k$ are derived from α and $\gamma=(\gamma_1,...,\gamma_k)$ by the equalities $\alpha_i = \alpha\gamma_i$ (cf. 6.14).

In some cases *X* might find it too difficult to specify $E[G(\mathbf{q})]$, whereas he might find it easier to specify his expected distance $E[D_2(\gamma,\mathbf{q})]$ between his prior vector γ and the true value of $\mathbf{q}$. Therefore, given the equality $E[D_2(\gamma,\mathbf{q})] = \text{VAR}(\mathbf{q}) = G(\gamma)/(\alpha+1)$ (see formulae (4.5) and (6.23)), the value of α can be found using the following formula:

$$(2) \qquad \alpha = \frac{G(\gamma) - E[D_2(\gamma,\mathbf{q})]}{E[D_2(\gamma,\mathbf{q})]}$$

The above interviewing procedures can be called *direct methods* since the assessment of a prior distribution is based on the direct specification of a sufficient number of its typical values (mean, variance, etc.). However, often direct methods fail since, as Winkler (1980, p. 95) remarks:

... it is often quite difficult for experts, even those with a good background in probability and statistics, to think about the model parameters and to quantify their prior information in terms of a probability distribution for these parameters.

In such cases Winkler (*ibid.*) suggests employing

... an *indirect method* whereby the expert assesses probability distributions for statistics based on samples from the process, with no mention being made of parameters. These distributions, which are called *predictive distributions*, involve observable variables instead of unobservable parameters. Hence, they should relate more directly to the expert's knowledge and experience and should be easier for the expert to think about than are prior distributions of parameters. [Author's italics].

For example, consider the prior predictive probability $p(t_n)$ of the occurrence of t_n Q_1-outcomes in a sequence of n trials of a given Bernoulli process *Ex*, where $0 \leq t_n \leq n$. It is known that the parameters α_1, α_2 of the Beta prior Beta(α_1,α_2) can be determined by fixing - for *two* different couples (n,t_n) - the predictive probabilities $p(t_n)$ derived from Beta(α_1,α_2) (cf. Winkler, 1972, pp. 206-208). If X specifies more than two predictive probabilities $p(t_n)$, then the result would as a rule be 'overdetermined': in other words, there would be no Beta(α_1,α_2) which corresponds *perfectly* to X's predictive probabilities. However, it would be erratic to discard any part (which part?) of the available information about X's prior opinions, especially when X has specified several predictive probabilities. The following appears to be a wiser strategy tc adopt:

(i) ascertain whether there are any Beta distributions where the derived predictive probabilities display a sufficient 'proximity' to the set of X's predictive probabilities,

(ii) if such distributions exist, select the one which provides the best fit (cf. Winkler, 1972, pp. 190-191; 1980, pp. 102-103).

The subjective approach can be seen as a sort of 'anarchic Bayesianism' where *every prior goes*. To reject the subjective approach it must be shown that, at least in certain situations, some degree of rational consensus on prior probabilities is attainable by applying appropriate normative principles (to be added to the probability axioms). A radical formulation of this claim is given in the aprioristic approach to BS.

7.2 THE CHOICE OF PRIORS IN BAYESIAN STATISTICS: THE APRIORISTIC APPROACH

The case of complete ignorance

The form of subjective Bayesianism as described above might be called *radical subjectivism*, given its refusal of *any* principle of rationality other than the probability axioms. This view may be seen as one of the two poles of a continuum of possible attitudes concerning the choice of priors, where the other pole - which may be called *radical objectivism* - is the view that, for any scientific problem, there is always a *unique* admissible prior. Between radical subjectivism and radical objectivism one finds different intermediate '*degrees of objectivism*' depending on the extent to which the accepted principles of rationality restrict the range of allowed freedom in the choice of prior distributions.

Although objective Bayesianism is a very old view going back to at least Laplace, it has been revitalized over the last fifty years by the works of H. Jeffreys and E. T. Jaynes. As Seidenfeld (1979a, p. 416) points out,

> ... objectivism (as defended by Jeffreys and Jaynes) ... argues for a *uniquely* admissible probability function ... given a knowledge state *K*. The probability function is *objective* because it is agent-invariant once the knowledge base *K* is fixed. Of course, objective Bayesianism requires extra inductive postulates [besides the probability axioms] to identify the uniquely admissible probability function.

While Jeffreys and Jaynes have tried to identify the uniquely admissible prior distribution for *any* kind of "knowledge state", the first supporters of objective Bayesianism - beginning with Bayes and Laplace - dealt only with states of *complete ignorance*.

As is known Bayes and Laplace thought that uniform priors were the uniquely admissible priors in the case where somebody is completely ignorant about the value of a parameter. Laplace justified this claim by his famous *principle of insufficient reason* (*PIR*) which may be stated as follows (Jeffreys, 1961, p. 33):

If there is no reason to believe one hypothesis rather than another, the probabilities are equal. ... *to say that the probabilities are equal is a precise way to say that we have no ground for choosing between the alternatives.*

A very common criticism of (PIR) is that very often the application of (PIR) to different 'partitions' of the same set of alternatives leads to mutually inconsistent probability assignments (cf. Seidenfeld, 1979b, pp. 11-12). The most striking cases of inconsistency are given by the so-called *paradoxes of re-parametrization* where a uniform prior on a continuous parameter θ determines non-uniform priors on certain parameters, such as θ^2, log θ, $1/\theta$ derived from the re-parametrization of θ.[3] This signifies that a uniform distribution on θ is incompatible, for instance, with a uniform distribution on $\phi = \theta^2$.

Now, assuming that a state of complete ignorance is represented by uniform priors - as (PIR) requires - we are then led to the conclusion that a person completely ignorant of the value of θ is not completely ignorant of the value of $\phi = \theta^2$ (and vice versa). Therefore one might be completely ignorant about the length of a square but quite well informed about its area. This 'arbitrariness of parametrization' conflicts with the presystematic intuition according to which the '*ignorance distributions*' representing a complete ignorance of a certain parameter should be invariant - in an appropriately defined sense of the term - relative to certain 'admissible' transformations of the parameter under investigation.

With the rejection of (PIR) new procedures for the formal representation of complete ignorance were designed to escape the paradoxes of re-parametrization. For instance, an important objective of Jeffreys's research program is to define ignorance distributions (or, equivalently: *informationless distributions*, *indifference distributions*) such that certain '*invariance properties*' are satisfied.[4] According to Jeffreys (1961, p. 119):

If we have no information relevant to the actual value of a parameter, the probability must be chosen so as to express the fact that we have none. It must say nothing about the value of the parameter, except the bare fact that it may possibly, by its very nature, be restricted to lie within definite limits.

Jeffreys's proposal is based on the assumption that the selected ignorance prior should respect certain invariance properties depending on the mathematical form of θ. For instance, according to Jeffreys (1961, p. 117):

(3) (a) If $-\infty < \theta < \infty$ then the ignorance prior is the (improper) uniform prior.

(b) If $0 \leq \boldsymbol{\theta} < \infty$ then the ignorance prior is the (improper) prior proportional to $1/\theta$.

The adoption of these priors is prompted by their invariance over important families of re-parametrizations. In particular, the priors in (a) and (b) are supported by the following invariance properties, respectively:

(a)' Given a parameter $\boldsymbol{\theta}$, where $-\infty < \boldsymbol{\theta} < \infty$, the uniform prior on $\boldsymbol{\theta}$ leads to priors of the same kind on any parameter obtained through a linear transformation of $\boldsymbol{\theta}$.

(b)' Given a parameter $\boldsymbol{\theta}$, where $0 \leq \boldsymbol{\theta} < \infty$, the prior proportional to $1/\theta$ leads to priors of the same kind on any parameter $\boldsymbol{\theta}^t$, with $t \neq 0$.

It follows from (a)' and (b)' that, within a given group of 'admissible' transformations of the parameter $\boldsymbol{\theta}$, the ignorance prior on one of the members of the group never leads to a non-ignorance prior on some of the other members. Hence, Jeffreys's approach succeeds in avoiding the paradoxes of re-parametrization arising in relation to (PIR).

On the other hand, serious criticisms have also been made against Jeffreys's standpoint. For instance, Barnett (1974, p. 179) asks:

Why should we use a concept of linear invariance when $[\mathbf{R}(\boldsymbol{\theta}) = (-\infty,\infty)]$, and power-law invariance when $[\mathbf{R}(\boldsymbol{\theta}) = (0,\infty)]$?

In general, it seems problematic to justify the choice of the specific group of transformations w.r.t. which the invariance of ignorance priors is requested. An analogous point is made by Fine (1973, p. 171):

A source of difficulty with the invariance approach ... is the necessity for the identification of the transformation groups that properly characterize our incomplete prior knowledge. ... On the one hand, a cautious selection of ... [the group of transformations] may not lead to a unique prior distribution ... On the other hand, it is often the case that the use of invariance so overdetermines the prior distribution that there does not exist an invariant distribution.

Apart from the mathematical problems relating to the under- and over-determination of priors, a more substantial question concerns the 'naturalness' of the criteria to be used for selecting the group of admissible transformations.[5] Invariance properties, indeed, are nothing but 'formal requirements of consistency' which are imposed on ignorance priors. In themselves they do not supply a formal representation of the intuitive concept of complete ignorance. As Huzurbazar (1980, p. 446) remarks:

... mere logical consistency, though essential, is not enough. It is also desirable that an invariance rule should lead to results agreeing with common sense.[6]

Although many 'indifference procedures' have been proposed in the statistical literature,[7] a satisfying explication of the notion of complete ignorance presumably has still to be found (cf. Barnett, 1974, p. 180).

Jeffreys (1948) himself, not being completely satisfied with his earliest invariance theory, proposed a new theory which is called the *theory of Invariants*.[8] As Seidenfeld (1979a, p. 418) remarks:

The innovation was to consider not just the quantity of interest, *m*, but also to consider the statistical distribution of the observable random variable which is to provide the information about *m*. That is, to specify an 'ignorance' distribution about *m* one must take into account the statistical model for the data which are to be the evidence acquired.

The theory of Invariants is much more powerful than Jeffreys's previous invariance theory and, in particular, allows identification of the ignorance prior for different kinds of parameters with a finite range of values. For instance, in the case of the parameter vector $\mathbf{q}$ of a multivariate Bernoulli process, the informationless prior derived from the theory of Invariants is the Dirichlet distribution Dir(1/2,...,1/2) (cf. Good, 1965, pp. 28-29).

However, it is not clear whether the new theory is more intuitively plausible than the old one. For instance, Lindley (1972, p. 71) remarks that:

A ... serious objection ... lies in the fact that the ignorance prior depends on the likelihood function. ... Why should one's knowledge, or ignorance, of a quantity depend on the experiment being used to determine it?[9]

The case of partial information: the maximum entropy approach

Although scientists are almost never in a state of complete ignorance, the identification of informationless priors for epistemic states of complete ignorance is important for other reasons. In fact, even if the actual epistemic state of a researcher includes much 'private information' about the parameter to be investigated, he may decide to report his experimental findings by using an indifference prior representing a supposed state of complete ignorance. When used in this function a prior is called *reference prior* (cf. Box and Tiao, 1973, p. 22).[10]

Having recognized the methodological importance of the case of complete ignorance, one has to recall that typically the researcher's initial cognitive state concerning the parameter $\boldsymbol{\theta}$ of a given process *Ex* is one of *partial information*. More specifically, his background knowledge *BK* would include some prior information *I* 'concerning' $\boldsymbol{\theta}$ in *Ex*. For instance, *BK* may include some theoretical information about certain relations occurring between $\boldsymbol{\theta}$ and other parameters characterizing *Ex*, or empirical information about the value of $\boldsymbol{\theta}$ in processes 'sufficiently similar' to *Ex*.[11]

Objective Bayesians assume that the above mentioned prior information *I* should not be neglected when selecting the prior distribution on $\boldsymbol{\theta}$. In particular Jaynes makes the stronger assumption that, starting from *I*, one can determine - by using suitable principles of rationality - a uniquely admissible 'objective prior' *cdf* $F^*(\theta)$ on $\boldsymbol{\theta}$ which will be referred to as *the partial information distribution* - or *the PI-distribution* - on $\boldsymbol{\theta}$, relative to *I*. According to Jaynes (1984, pp. 167-168):

> The direct problem is: Given some verbal statement *I*, which says something relevant to $\boldsymbol{\theta}$ without uniquely fixing it, translate *I* into a prior probability assignment $f(\theta)$.

Jaynes's analysis of this problem may be illustrated by the following example. Consider a discrete parameter $\boldsymbol{\theta}$ where the possible values are $a_1,...,a_t$. Given a discrete probability distribution $p=(p_1,...,p_t)$ on $\boldsymbol{\theta}$ - where $p_i = p(a_i)$ - then the *entropy* $H(p)$ of p is defined as follows:

$$(4) \qquad H(p) = -\Sigma\, p_i \log(p_i)$$

The entropy $H(p)$ introduced by Shannon (1948), is commonly interpreted as a measure of the *uncertainty* - or lack of information - of a subject whose probability distribution on $\boldsymbol{\theta}$ is p. This interpretation of $H(p)$ is also accepted by Jaynes, who suggests the following *principle of maximum entropy (PME)* for selecting the objective prior distribution p^*:

(PME) (a) Identify the set $\mathbf{p}_I$ of all probability distributions on $\boldsymbol{\theta}$ 'consistent' with the prior information *I*;

(b) then select within $\mathbf{p}_I$ the *maximum entropy distribution* p^* maximizing $H(p^*)$.

Subsequently, p^* will be referred to as the *MAXENT distribution* (cf. Seidenfeld, 1986), and "*MAXENT*" will denote the forms of objective Bayesianism using (PME) or other 'entropic principles'.

Some explicative remarks about the 'consistency clause' (a) of (PME) are necessary. In many applications of MAXENT, the partial information I includes a number of 'prior constraints' $c_1,\ldots,c_m$ given by the expected values of certain functions $f_1(\boldsymbol{\theta}),\ldots,f_m(\boldsymbol{\theta})$ of $\boldsymbol{\theta}$ (cf. Jaynes, 1979, pp. 45-47 and Rosenkrantz, 1977, pp. 55-56). Prior constraints of this sort - such as mean, variance and other moments of $\boldsymbol{\theta}$ - will be referred to as *moment constraints* or *m-constraints*. However, many other kinds of *distributional constraints* on $p(\theta)$ can be used: in principle one can employ any constraint leading to some restriction on the set of possible prior distributions on $\boldsymbol{\theta}$ (cf. Fine, 1973, p. 172). One might distinguish between (a) '*propositional constraints*', which are expressed by a certain proposition about the value of the investigated parameter $\boldsymbol{\theta}$ as, for instance: "The value of $\boldsymbol{\theta}$ is lower than 2", and (b) '*probabilistic constraints*' consisting in the specification of some features - in general *typical values* as moments, fractiles, etc. - of the prior probability distribution on $\boldsymbol{\theta}$ as, for instance: "There is at least a 90% probability that $\boldsymbol{\theta}$ is less than 2" and "The mean value of $\boldsymbol{\theta}$ is 2" (cf. Jaynes, *ibid.*, pp. 44-45 and Seidenfeld, 1986, p. 471).

An attractive feature of MAXENT is the existence of a general solution to the problem of finding the MAXENT distribution in cases of this kind (cf. Jaynes, *ibid.* pp. 46-47).

A concept of entropy analogous to (4) can be defined also for a *pdf* $f(\theta)$ on a continuous parameter $\boldsymbol{\theta}$:

$$(5) \qquad H[f(\theta)] = -\int f(\theta) \log[f(\theta)]\, d\theta$$

Unfortunately, if applied to a continuous parameter $\boldsymbol{\theta}$, (PME) gives rise to inconsistencies not that different from the paradoxes of re-parametrization generated by (PIR), in the sense that the maximally entropic *pdf* $f^*(\theta)$ for $\boldsymbol{\theta}$ is typically inconsistent with the maximally entropic distributions for other parameters $\boldsymbol{\phi}$ obtained by certain 'smooth' transformations of $\boldsymbol{\theta}$ (cf. Seidenfeld, 1979a, p. 424; 1986, p. 475).

The solution suggested within MAXENT employs the notion of *cross entropy (divergence)*, also known as *Kullback-information* (Kullback, 1959). In the continuous case the cross entropy $I_K(f_2,f_1)$ between two *pdf*'s $f_2(\theta)$ and $f_1(\theta)$ is defined as follows:

$$(6) \qquad I_K(f_2,f_1) = \int f_2(\theta) \log[(f_2(\theta)/f_1(\theta)]\, d\theta$$

$I_K(f_2, f_1)$ is commonly interpreted as a measure of the *distance* - or *dissimilarity - between* $f_2(\theta)$ and $f_1(\theta)$ since, in one sense, it represents the 'change of belief' involved in shifting from $f_1(\theta)$ to $f_2(\theta)$.[12]

Let $f_0(\theta)$ be the informationless prior on $\boldsymbol{\theta}$ derived from an appropriate invariance theory. Then, within MAXENT the PI-distribution $f^*(\theta)$ is selected by applying the rule of *minimization of the change of belief (MCB)* which can be stated as follows:

(MCB) (a) Identify the set $\mathbf{f}_I$ of all *pdf*'s $f(\theta)$ on $\boldsymbol{\theta}$ 'consistent' with the prior information I;

(b) select within $\mathbf{f}_I$ the *pdf* $f^*(\theta)$ which minimizes the cross entropy divergence $I_K(f^*, f_0)$.

According to (MCB) the 'new' cognitive state represented by $f^*(\theta)$ should be - compatibly with I - '*maximally similar*' to the 'old' cognitive state represented by $f_0(\theta)$.[13]

A positive feature of (MCB) concerns its application to discrete parameters. In fact it can be proved that, if one assumes, as is usually the case, that the ignorance distribution p_0 for a discrete parameter $\boldsymbol{\theta}$ is the uniform distribution, then the PI-distribution p^* minimizing $I_K(p^*, p_0)$ - within the set $\mathbf{p}_I$ of all the distributions on $\boldsymbol{\theta}$ 'consistent' with I - is identical to the PI-distribution which is derived from (PME) (cf. Rosenkrantz, 1981, Chapter 4.3, p. 3 and Seidenfeld 1986, p. 474). Therefore, in the case of discrete parameters, (MCB) is logically equivalent to the principle of maximum entropy (PME). Since (MCB) can be seen as a generalization of (PME) to continuous parameters, then also the PI-distributions selected on the basis of (MCB) will be called MAXENT-distributions.

A critical evaluation of MAXENT

Some important problems concern the 'nature' and the 'source' of the constraints included in I. Regarding the first problem Jaynes (1978, p. 61) points out that:

> In our mathematical formalism, a "constraint" is some piece of *information* that leads to modify a probability distribution ... It is perhaps not yet clear just what we mean by "constraints" in a physical experiment. Of course, by these we do not mean the gross constraining linkages by levers, cables, and gears of a mechanics textbook, but something more subtle. In our applications, a "physical constraint" is any physical influence that exerts a systematic tendency - however slight - on the outcome of an experiment.

Jaynes (*ibid.*, pp. 62-72), for instance, considers the data collected by the Zurich astronomer R. Wolf who, in the period roughly between 1850-1890, tossed an ivory die 20,000 times. After remarking that "any imperfection in the die may ... give rise to a 'physical constraint'" (*ibid.*, p. 62), Jaynes (*ibid.*, p. 63) then considers two typical imperfections of a die:

> The most obvious imperfection is that different faces have different numbers of spots. This affects the center of gravity, because the weight of ivory removed from a spot is obviously not (in any die I have seen) compensated by the paint then applied. Now the numbers of spots on opposite faces add up to seven. Thus the center of gravity is moved toward the "3" face, away from "4", by a small distance ε corresponding to the one spot discrepancy. The effect of this must be a slight frequency difference which is surely, for very small ε, proportional to ε.

The (estimated) value of this "slight frequency difference" will be one of the distributional constraints to be used when applying (PME) to Wolf's Dice.

It is difficult to imagine that the probabilistic constraints in I are 'given directly' to us as a result of some previous experimental evidence or accepted theory. It seems more reasonable to assume that there is some 'prior inference' leading from the background knowledge BK to I. A possible answer to the question about the *source* of the constraints included in I is suggested by the above example of Wolf's Dice Data. It suggests, indeed, that the conjectures about the "physical constraints" - or "physical influences" - working in the investigated experimental set-up derive from a certain background knowledge or, more specifically, from the laws of classical mechanics together with the observed physical features of the dice used (and of other dice similar to them).

More generally, it seems to me that, in the applications of MAXENT commonly considered in the literature, the constraints in I may be interpreted as 'formal representations' of certain *guesses* - derived from the available background knowledge BK - concerning a corresponding set of *physical constraints*.[14] Let us consider, for instance, the probabilistic constraints $c_1,...,c_m$ specifying some moments - or other typical values - of certain functions $f_1(\boldsymbol{\theta}),...,f_m(\boldsymbol{\theta})$ of the parameter $\boldsymbol{\theta}$ of a given process Ex. Here $f_1(\boldsymbol{\theta}),...,$ $f_m(\boldsymbol{\theta})$ denote objective features of Ex, while $c_1,...,c_m$ can be seen as the formal representations of certain well-grounded conjectures - derived BK - about $f_1(\boldsymbol{\theta}),...,f_m(\boldsymbol{\theta})$. For instance, if we guess - starting from BK - that the value of $f_1(\boldsymbol{\theta})$ is a, we may formally represent this guess by the constraint c_1: "the expected value of $f_1(\boldsymbol{\theta})$ is a".

Now the question concerning the source of the probabilistic constraints $c_1,...,c_m$ included in a given prior information can be put as follows: by which 'prior inferences' does *BK* - which typically consists of experimental and theoretical propositions - lead to the 'acceptance' of $c_1,...,c_m$? In other words: which is the path leading from *BK* to certain probabilistic constraints on $\mathbf{\theta}$?

It seems to me that the claim that the MAXENT distribution is 'the objective prior' involves, at least implicitly, the assumption that the distributional constraints in *I*, far from expressing simple subjective opinions, are based on some 'objective' - and intersubjectively shared - background knowledge *BK*. Unfortunately, in the literature on MAXENT one does not find any general analysis of the prior inferences leading from *BK* to the distributional constraints in *I* and, at present, it is not clear whether a satisfying general account of these inferences is indeed possible (cf. final section of this chapter).

Many objections to MAXENT concern the cognitive status and the intuitive plausibility of the principles (PME) and (MCB).[15] For instance, Miller (1980, p. 264) points out that,

> ... even if a unique proper prior distribution were always attainable by use of some variant or other of Jaynes' rule, it would still be pertinent to ask why that prior should be used. Uniqueness is not itself any guarantee of objectivity, or rather, the objectivity that it does guarantee has no obvious connection with correctness.

Some reasons for the assumed objectivity - and universal applicability - of the MAXENT distributions derived from (PME) and (MCB) should be given.

Consider, for instance, (MCB). Arguing for the universal applicability of (MCB) involves the justification of two claims: (i) that within the set of all the distributions satisfying the constraints in *I*, the MAXENT distribution $f^*(\theta)$ should be selected in any case so as to minimize the distance between $f^*(\theta)$ and the ignorance distribution $f_0(\theta)$ and (ii) that the 'right' measure of distance between two probability distributions is given in any case by the cross entropy divergence (see (6)).

Even supposing that claim (i) may be justified,[16] serious doubts arise in relation to claim (ii). For instance: since in statistical literature there are so many measures of dissimilarity (distance) between probability distributions, why should we choose just the cross entropy divergence as the 'universally appropriate' distance measure? One should recall, in this connection, that the definition of plausible measures of distance for *epistemic* probability functions is still an open question and that cross entropy divergence is no more than a possible solution to this problem.[17]

Further, it should be recalled that, generally speaking, the (dis)similarity between two 'complex entities' is a 'multi-dimensional matter' and, therefore, a 'global measure' of (dis)similarity has to reflect the relative importance attributed to several 'partial similarities' between different features of the entities in consideration. Since probability distributions are also somewhat complex entities, it is reasonable to assume that there is no 'universally appropriate' measure of dissimilarity (distance) between epistemic probability distributions; on the contrary, it would appear that different distance measures may be used in different cognitive contexts. This is tantamount to rejecting claim (ii).

Analogous remarks apply, in the case of discrete parameters, to the principle (PME) which requires maximization of the uncertainty of the prior distribution as defined by its Shannon entropy. In fact, although accepting the idea that in the selection of prior probabilities uncertainty should be maximized, one might wonder whether $H(p)$ is a universally appropriate explicatum for the uncertainty of a probability distribution p. The introduction of many alternative indices of entropy (uncertainty) - for instance, the "entropy of order a" proposed by Rényi (1961), the "weighted entropy" proposed by Guiasu and many others (see Guiasu, 1977, Chapters 4-5) - seems to suggest that this is not the case. It would appear much more plausible to suppose that different measures of uncertainty may be appropriate in different cognitive contexts. This conflicts with the strong belief held by Jaynes that (PME) is universally applicable to all discrete parameters.

7.3 THE SUBJECTIVE INTERPRETATION OF THE THEORY OF INDUCTIVE PROBABILITIES

There are two basic assumptions underlying the subjective interpretation of TIP:

(i) the requirements of adequacy used in TIP should be *descriptive principles* able to capture important features of the inductive probabilities of the researchers engaged in certain types of empirical research;

(ii) such principles should leave room for a wide variety of inductive methods.

According to Jeffrey (1973, pp. 302-303) Carnap's approach to TIP can be seen as a particular form of subjectivism:

> ... perhaps [Carnap] is describable as a special sort of subjectivist. Differences among the [C-systems] which represent our various inductive intuitions are matters of individual

psychology, while the respects in which all of our [C-systems] agree represent human nature - general psychology. Perhaps Carnap would accept this subjectivist reading, but would think it appropriate to dignify with the name 'logical' the common features of all [C-systems] ...

Inductive logic is a *description* of the norms we shall find ourselves accepting, once we have thought the matter through. It is no part of Carnap's program, to base those norms on still deeper philosophical considerations.

This is in keeping with the first of the two above assumptions. Regarding the second assumption, Jeffrey (*ibid.*, pp. 302) points out that:

Carnap was prepared to admit the possibility that different people might have somewhat different inductive intuitions, e.g. when, ca. 1951, he thought that the right [C-system] might be found somewhere in the continuum of inductive methods, he thought that different people might discover that they had somewhat different values of λ ... He thought it possible that these differences were irreducible, so that ... there might be no such thing as *the* [C-system] which represents *our* inductive intuitions. Rather there might prove to be a range of inductive intuitions among us, which might prove to be represented by different [C-systems].

For most of his thirty five years of work on TIP, Carnap accepted the idea that TIP leaves researchers considerably free to choose their prior probabilities. According to Carnap (1980, p. 106):

The person *X* wishes to assign rational credence values to unknown propositions on the basis of the observations he has made. It is the purpose of inductive logic to help him to do this in a rational way; or, more precisely, to give him some general rules, each of which warns him against certain unreasonable steps. The rules do not, in general, lead him to specific values; they leave some freedom for choice within certain limits.

Referring to GC-systems, Carnap (1980, pp. 111-112) states that,

... although certain boundaries for λ can be determined objectively by considerations of rationality requirements, within these limits everyone is free to make his choice as he pleases. We might call this a (modified) *subjectivist point of view*. (A similar view, but without acceptance of objective limits might be called a purely subjectivist view.)

Regarding the choice of a λ-value by an individual *X*, Carnap (*ibid.* p. 112-113) points out that:

... we might regard *X*'s choice of a λ-value ... as determined by, and therefore symptomatic of, certain features of *X*'s personality. ... *X*'s choice should not be regarded as arbitrary or unpredictable in the sense that a person of *X*'s nature could just as well have made another choice; *X*'s choice is rather regarded as determined or at least influenced by his personality.

Lastly, Carnap (*ibid.*, pp. 113-114), referring to the different effects of experimental evidence on the posterior probabilities depending on the value of λ, suggests that,

... λ measures the resistance or inertia against this influence of experience. ... from the personalist point of view it is regarded as possible that rational people use different λ-values ... The difference may be attributed to their different inductive inertia.[18]

It is clear that according to the subjective interpretation of TIP, the *optimum inductive method* is the one which supplies the best formal representation of the researcher's beliefs. To identify such a method, procedures similar to those illustrated in the subjective approach to the choice of priors in BS should be used.

7.4 THE APRIORISTIC INTERPRETATION OF THE THEORY OF INDUCTIVE PROBABILITIES

There are two basic assumptions underlying the aprioristic interpretation of TIP:

(i) the requirements of adequacy used in TIP should be *a priori normative principles*, i.e., principles of rationality justified by a priori considerations;

(ii) on the basis of such principles, a unique inductive method should be specified or, at least, the range of admissible inductive methods should be considerably restricted.

Both assumptions were advocated by Carnap who, during his long career as an inductive logician, often oscillated between subjectivism and apriorism. For instance, Carnap (1968, p. 265) stresses the importance of "inductive intuition" in justifying the axioms for TIP:

I think that it is not only legitimate to appeal to inductive reasoning in defending inductive reasoning, but that it is indispensable. ... If a person were unable to distinguish valid from invalid steps in inductive reasoning, even in the simplest cases, then it would be hopeless to try to convince him of anything in inductive logic. In order to learn inductive reasoning, we must have what I call the ability of *inductive intuition*.

Although Carnap is somewhat vague about the relationships between inductive axioms and inductive intuitions, it seems to me that he does not consider inductive axioms as a mere *description* of our inductive intuition but that he looks at inductive intuition as a source of a priori arguments for the *justification* of inductive axioms.[19]

Regarding the second assumption, (Carnap, 1968, p. 314) points out that he is inclined

> to base the choice of an inductive method, and in particular the choice of a value of λ, ... only on a priori considerations. The considerable narrowing of the range of admissible λ-values which we have achieved in the course of the last ten years was obtained exclusively by *a priori* arguments.

More specifically, let us examine how, from Carnap's (1971, 1980) aprioristic standpoint, the decision to use a given GC-system is justified. This decision is seen as a 'four step decision' (see Carnap, 1980, p. 106):

(i) the decision to use a GC-system or not,
(ii) the choice of the γ-values, or prior predictive probabilities, $\gamma_1,\dots,\gamma_k$,
(iii) the determination of the range of admissible λ-values,
(iv) the choice of a specific λ-value.

Before considering these steps in detail, let us recall that, w.r.t. an arbitrary family of properties $\mathbf{Q}=(Q_1,\dots,Q_k)$, Carnap (1971, §§ 2 and 4-5) assumes that:

(a) the properties, or "*attributes*", $Q_1,\dots,Q_k$ can be represented as extended regions of an "*attribute space*" related to **Q**,
(b) for any attribute space a suitable metric (i.e., a distance function based on similarity relations) can be defined,
(c) starting from this metric, the "*width*" of each attribute and the "*distances*" between couples of attributes can be determined.

The basic assumption underlying Carnap's aprioristic approach is that the assignment of inductive probabilities "to propositions involving attributes of a given family is frequently influenced by certain features of these attributes and relations among them" (Carnap, 1980, p. 8) such as the above mentioned widths and distances.

More specifically, w.r.t. step (i), Carnap (1980, p. 107) maintains that GC-systems should be used (1) whenever **Q** is an "*equal-distance family*" where the distances between the different couples of properties are (approximately) equal to each other, and (2) whenever **Q** is a "*long-distance* *family*", where none of the distances between pairs of properties is below a certain threshold value. This claim is defended as follows.

According to Carnap (1980, p. 42) when **Q** is an equal-distance (long-distance) family, any 'analogy by similarity' between different properties in **Q** should be excluded and accordingly the inductive probabilities should satisfy the so-called η-equality (cf. (6.6)). It can be proved that, when $k > 2$, the η-equality is entailed by the requirement of restricted relevance (RR) (Carnap, *ibid.*, p. 87, T19-1). Referring to this logical relation between η-equality and (RR), Carnap (*ibid.*, pp. 104-105) argues that, if η-equality is assumed, (RR) should be adopted[20] and therefore GC-systems should be used. (Remember that, when $k > 2$, GC-systems can be axiomatized using (RR), together with (Exc) and (IP): cf. (6.38).)

Regarding step (ii), Carnap (1980, pp. 33-34 and 106) suggests that the γ-value γ_i of each property Q_i should be proportional to the width of the corresponding attribute. For instance, the GC-systems "*with γ-equality*", or C-systems, should be used where all attributes have equal width (Carnap, pp. 106-107).

Concerning the determination of the range of admissible λ-values for GC-systems (step (iii)), Carnap (1980, pp. 107-111) argues that, in the particular case of C-systems, this range may be considerably restricted given the following a priori considerations.

Consider two complete descriptions e_1 and e_2 of a sample of n individuals (with $n > k$) where e_1 is minimally ordered (i.e., the k numbers $n_1,...,n_k$ differ at most by one) and where e_2 is maximally ordered (i.e., all the members of the sample have the same property). Having proved that in any C-system $p(e_2) > p(e_1)$, Carnap (*ibid.* pp. 108-109) proposes a "*uniformity requirement*" (UR) which strengthens the above inequality by replacing e_1 and e_2 with the corresponding sets $\ddot{e}_1$ and $\ddot{e}_2$ (see Chapter 2.3):

(UR) (a) $p(\ddot{e}_2) \geq p(\ddot{e}_1)$
(b) $p(\ddot{e}_2) \neq p(\ddot{e}_1)$

The intuitive idea underlying (UR) is that a maximally ordered sample has a "preferential status" and - consequently - should receive "preferential treatment" w.r.t. a minimally ordered sample (Carnap, *ibid.*, p. 109).

Carnap (*ibid.*, pp. 109-110) proves that accepting clause (a) is tantamount to applying - for any family $(Q_1,...,Q_k)$ where $k \geq 2$ - the restriction "$\lambda \leq k$" on the range of admissible λ-values, while the stronger restriction "$\lambda < k$" should be applied if clause (b) is also accepted.

Other a priori considerations lead Carnap (*ibid.*, pp. 110-111) to exclude "very small" values of λ, so as to reach the provisional conclusion that the range of admissible λ-values is given by $1/2 < \lambda < k$.

The four step selection of the optimum inductive method is completed, in step (iv), by the choice of a specific λ-value (see Carnap 1980, pp. 111-118). Concluding a detailed analysis, Carnap (*ibid.*, p. 118) tentatively suggests that "all the long-distance families, for any attribute space and for any size of the family [i.e., for any value of *k*], have practically the same λ-value". In combination with (UR)(a), this proposal implies that $\lambda \leq 2$. Restricting, for simplicity, the choice of λ to integer values, this is equal to assuming $\lambda = 1$ and $\lambda = 2$ as the only admissible values of λ. Finally, if (UR)(b) is also accepted, one is led to selecting $\lambda = 1$ as the uniquely admissible value of λ. Indeed, Carnap (*ibid.*, p. 119) expresses his preference for $\lambda = 1$ as the *optimum λ-value* for all the long-distance families.[21]

Some remarks on Carnap's aprioristic interpretation of TIP may be useful to note. For instance, Carnap's justification of (RR) does take account, to some extent, of the cognitive context of a given empirical research, since the metrical structure of (the attribute space related to) the family **Q** used to describe the population or process under investigation is one of the features of this context. Notwithstanding this, Carnap's justification is highly aprioristic since he claims that (RR) should be adopted *wherever* an equal-distance - or long-distance - family **Q** is employed, independently of the *specific* background knowledge *BK* available within a given cognitive context.

It seems to me that, against Carnap's claim, *BK* plays an important role in justifying the decision to adopt - or reject - the requirement of adequacy (RR). Let us suppose, for instance, that, according to the available background knowledge *BK* concerning the population of cars produced over the last year by a certain factory, all cars are coloured with one of the colours in the family **Q** = (Yellow, Orange, Black). According to Carnap's view, (RR) should *never* be applied in the case where **Q** is used since, due to the nearness between Yellow and Orange, **Q** is neither an equal-distance family nor a long-distance family. However, it seems to me that Carnap is wrong on this point, since there may be cases where the available *BK* justifies the adoption of (RR) also where the population under investigation is described using **Q**. For instance, in the case considered above, I believe that, if the available *BK* concerning the marketing strategies of factories excludes that they have the tendency to concentrate their production of cars on chromatically similar kinds, then (RR) should be adopted. (For a more detailed 'contextual justification' of the decision to adopt inductive methods satisfying (RR), see Section 6).

Now consider a case where, according to the available *BK*, certain widely adopted marketing strategies suggest the large scale production of some sharply differentiated 'basic kinds' of cars and a limited production of some 'variants' of each basic kind. This implies that, typically, a factory does not engage in large scale production of balls of very similar colour. Hence, as far as **Q** = (Yellow, Orange, Black) is concerned, there should be a strong 'negative correlation' between the proportions of Yellow and Orange balls within the set of cars produced last year by a factory. It follows that, in spite of the similarity between Yellow and Orange, there should be a 'negative (inductive) analogy' between these two properties, in the sense that the observation of a yellow ball should be ***negatively relevant*** to the hypothesis that the next observed ball will be orange.[22] This conflicts with Carnap's view (1980, pp. 42-49) that a high proximity between two properties in a given attribute space should always imply a 'positive (inductive) analogy' between them. On the contrary, this example shows that the occurrence of a positive (negative) inductive analogy between two properties depends - rather than on their proximity in a certain attribute space - on the specific information conveyed by *BK*.[23]

One can imagine many other, more or less realistic, situations characterized by 'competition' between similar properties, so that the positive inductive analogy between two properties is inversely proportional to their proximity in a given attribute space. For instance, in a given ecological context, there may be strong competition between similar species. Another example concerns the competition among political parties where each party tends to increase its electorate at the expense of other ideologically similar parties.

Of course one can also imagine situations where the available *BK* implies that the positive inductive analogy between two properties is directly proportional to their proximity in a given attribute space. For instance, *BK* may imply that the 'causal mechanism' through which a given population is generated tends to produce positively correlated percentages of similar properties. However, there is no reason to believe that this kind of background knowledge is available in any cognitive context. Hence, it is difficult to see why one should accept Carnap's idea that a high proximity between two properties in a given attribute space should *always* imply a positive inductive analogy between them.

Analogous objections may be raised against Carnap's aprioristic view of the choice of γ-values (step (ii)): in Chapter 8.1 it is argued that this choice should be affected by the available *BK*.

Regarding Carnap's postulate (UR), I agree with Hilpinen (1973, p. 325) that:

The postulate ... expresses a strong *a priori* belief in a high degree of uniformity in our universe of individuals. It may be justified in special cases (for instance, on the basis of suitable background information), but it is hardly acceptable as a general principle of inductive logic.

Consider that the expected disorder $E[G(\mathbf{q})]$ for a GC-system (γ,λ) is a function of λ and the Gini diversity $G(\gamma)$ of the prior vector γ - i.e., $E[G(\mathbf{q})] = [\lambda/(\lambda + 1)]G(\gamma)$ - where $G(\gamma)$ is the maximum value of $E[G(\mathbf{q})]$ for a GC-system with prior vector γ (see (6.24), (6.25), and (6.33)). Taking account of this equality, Hilpinen's remark can be restated as given below.

Since (UR) implies that $\lambda < k$, it also implies that $E[G(\mathbf{q})] = [\lambda/(\lambda + 1)]G(\gamma) < [k/(k + 1)]G(\gamma)$, i.e., that $E[G(\mathbf{q})]$ is lower than $k/(k + 1)$ times the maximum value $G(\gamma)$. While this is a weak restriction for a family $\mathbf{Q}=\{Q_1,...,Q_k\}$ with a large k, it is quite strong for small values of k. For instance, for $k = 2$ the restriction imposed by (UR) is given by the inequality $E[G(\mathbf{q})] < (2/3)G(\gamma)$ which corresponds to "a strong *a priori* belief in a high degree of uniformity in our universe". However, it is difficult to see why this strong a priori belief should be adopted in *any* empirical research, independently of the population - or process - under investigation and the family $\mathbf{Q}=(Q_1,Q_2)$ used to describe it.[24]

It seems much more plausible to assume - as will be discussed in Chapter 8.2 - that a researcher's initial opinions about the degree of disorder in a population depend more on his 'contextual information' than on a priori assumptions.

7.5 THE CONTEXTUAL VIEW OF PRIOR PROBABILITIES

Some general considerations

We saw above (Section 2) that (partially) objective Bayesian approaches are characterized by the adoption of at least one principle of rationality other than the probability axioms, and that the *degree of objectivism* of a Bayesian approach is related to the 'strength' of the restrictions imposed by such principles of rationality on a researcher's freedom in his choice of prior probabilities.

Here I would like to make the distinction between two kinds of principles of rationality - or inductive principles - i.e., *universal* and *contextual principles*. *Universal principles* - or *a priori principles* - are claimed to hold in any cognitive context and, accordingly, are justified by a priori considerations. On the contrary, *contextual principles* are claimed to hold only

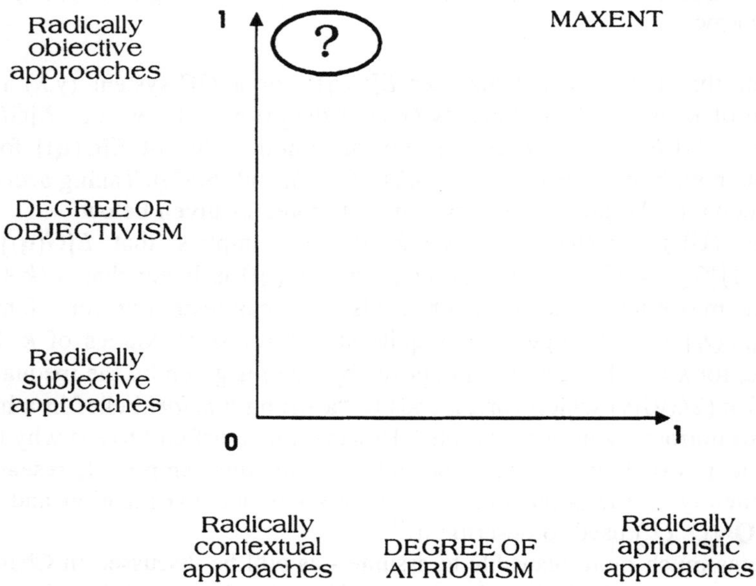

Figure 1

The 'two-dimensional space' of Bayesian approaches

in certain appropriate cognitive contexts and, accordingly, they are justified w.r.t. the particular kind of background knowledge which is available in such contexts.

Objective Bayesian approaches may be differentiated according to their *degree of apriorism* which depends on the number and the 'strength' of the adopted a priori principles. In particular, between the two poles represented by 'radically aprioristic approaches' and 'radically contextual approaches' one can visualize a wide range of 'partially contextual approaches' including both universal and contextual principles.

By combining the classification of Bayesian approaches according to their degree of objectivism with the classification according to their degree of objectivism one obtains the 'bidimensional space' of Bayesian approaches which is represented in Figure 1.

Notice that in Figure 1 the MAXENT approach is represented as a radically aprioristic and radically objective one. A reason frequently advocated for adopting MAXENT is, indeed, its radical objectivism, i.e., the fact that the a priori principles (PME) and (MCB) always allow identification, on the basis of the available partial information, of the uniquely admissible 'objective prior' in a given context. However, below I will argue that the need for objectivity (intersubjectivity) may also be satisfied by adopting a contextual approach. More specifically, I will show that (a) contextual principles may exist, and (b) that a contextual approach may be (radically) objective.

Regarding claim (a), it should be noted that a requirement of adequacy for a prior distribution may be seen as a particular type of distributional constraint. Let us consider, for instance, the parameter vectors $(\boldsymbol{\theta},\boldsymbol{\phi})$ and the probabilistic constraint c - where c = "$\boldsymbol{\theta}$ and $\boldsymbol{\phi}$ are independent" - on the prior distribution $p(\theta,\phi)$ (cf. Domotor, 1985. p. 75). One sees that c, unlike the constraints which are commonly used in the illustrations of MAXENT, does not specify any typical value of $p(\theta,\phi)$. On the contrary, c is a sort of 'formal constraint' expressing a *requirement of adequacy* which should be satisfied by the mathematical form of $p(\theta,\phi)$. To say that the requirement of adequacy c is a contextual principle is equal to saying that c may be derived from the 'objective' - and intersubjectively shared - background knowledge *BK* available in a given cognitive context. It seems to me that assuming this possibility is not more complicated, in principle, than assuming the possibility of a 'contextual justification' of the *m*-constraints and the other kinds of distributional constraints employed in the usual applications of MAXENT. A concrete example of this possibility will be given in Section 6, where I will show that the requirement of adequacy - or formal constraint - (CM), used for characterizing Dirichlet distributions (cf. Chapter 6.6), can be justified w.r.t. a certain background knowledge and, hence, can be seen as a contextual principle.

Concerning claim (b) - according to which a contextual approach may be (radically) objective - note that the use of (CM) in the axiomatization of the Dirichlet family for a parameter vector $\mathbf{q}=(\mathbf{q}_1,\ldots,\mathbf{q}_k)$ provides a concrete example of the possibility that a contextual requirement might impose a very strong restriction on the possible prior distributions on $\mathbf{q}$. Notice too that, given the formal constraint (CM) and other k distributional constraints (for instance the mean vector $\gamma=(\gamma_1,\ldots,\gamma_k)$ and the expected Gini diversity $E[G(\mathbf{q})]$), the uniquely admissible prior distribution on $\mathbf{q}$ for a given cognitive context can be identified (cf. (1)).[25] This signifies that the contextual Bayesian approach related to the contextual principle (CM) is radically objective.

Let us now return to the previously mentioned problem (Section 2) of providing an account of the 'prior inferences' from the available *BK* to certain *m(ean)*-constraints in the prior information *I*.

The basic idea underlying this account is that, in several cases, the prior beliefs of the researchers regarding parameter $\boldsymbol{\theta}$ of the experimental process *Ex* under investigation are (or, at least, should be) determined by certain information included in *BK* such as:

(i) all available *experimental information* on the value of $\boldsymbol{\theta}$ in processes 'sufficiently similar' to *Ex*;[26]

(ii) all available *theoretical information* on the relations between the value of $\boldsymbol{\theta}$ and the values of other quantities $\boldsymbol{\phi}_1,\ldots,\boldsymbol{\phi}_m$ in processes 'sufficiently similar' to *Ex* (together with information available on the values of $\boldsymbol{\phi}_1,\ldots,\boldsymbol{\phi}_m$ in *Ex*).

Such information may be called *external information* on $\boldsymbol{\theta}$ in *Ex*, or *EIn*($\boldsymbol{\theta}$,*Ex*).[27] In certain cognitive contexts the external information *EIn*($\boldsymbol{\theta}$,*Ex*) in *BK* may be 'rich' enough to suggest a 'prior point estimate' of $\boldsymbol{\theta}$ in *Ex*. This prior estimate will now be referred to as the *external estimate* of $\boldsymbol{\theta}$ and denoted by *Est*($\boldsymbol{\theta}$). More generally, in certain cognitive contexts *EIn*($\boldsymbol{\theta}$,*Ex*) may suggest an external estimate *Est*[$h(\boldsymbol{\theta})$] of a given function $h(\boldsymbol{\theta})$ of $\boldsymbol{\theta}$.

Although external estimates are made in a somewhat informal and spontaneous manner, it will be assumed that they satisfy the following minimal requirement of adequacy:

(Ext) (a) $Est(a\boldsymbol{\theta}_1 + b\boldsymbol{\theta}_2 + c) = aEst(\boldsymbol{\theta}) + bEst(\boldsymbol{\theta}_2) + c$

(b) If, for any $\theta \in \mathbf{R}(\boldsymbol{\theta})$, $\theta \geq 0$ then $Est(\boldsymbol{\theta}) \geq 0$.[28]

Note that, if a researcher is able to make two of the three estimates $Est(a\boldsymbol{\theta}_1+b\boldsymbol{\theta}_2+c)$, $Est(\boldsymbol{\theta}_1)$, $Est(\boldsymbol{\theta}_2)$ using appropriate 'direct procedures', he may derive the third using (Ext)(a) as an 'indirect procedure'.

In my view external estimates are the 'intermediate step' in the prior inference from *BK* to the *m*-constraints in *I*. More specifically, I would assume that the prior inferences from external estimates to *m*-constraints are ruled by the following *requirement of cognitive coherence (CC)*:

(CC) (a) Given the external estimate *Est*($\boldsymbol{\theta}$), the following *m*-constraint on $p(\theta)$ should be included in *I*: "$E(\boldsymbol{\theta}) = Est(\boldsymbol{\theta})$".

(b) Given the external estimate $Est[h(\boldsymbol{\theta})]$, the following *m*-constraint on $p(\theta)$ should be included in *I*: "$E[h(\boldsymbol{\theta})] = Est[h(\boldsymbol{\theta})]$".

The intuitive idea underlying (CC) is that our *m*-constraints on $p(\theta)$ should provide a formal representation of our (informal) prior estimates of $\boldsymbol{\theta}$ - or other parameters related to $\boldsymbol{\theta}$ - which are derived from the available background knowledge *BK*.

The above account of the prior inferences from *BK* to the *m*-constraints in *I* is represented in the Figure 2.

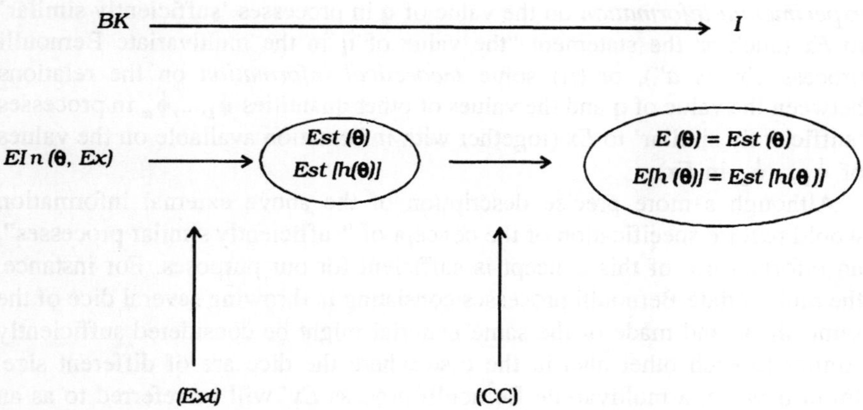

Figure 2

The prior inferences from the background knowledge BK to the m-constraints in I

7.6 A CONTEXTUAL JUSTIFICATION OF DIRICHLET DISTRIBUTIONS AND GC-SYSTEMS

In my opinion a context-dependent justification of the decision to use a prior Dirichlet distribution or, equivalently, a GC-system, can be given as shown below.

Suppose that the background knowledge *BK* available in a given cognitive context includes the assumption that the experimental process *Ex* - whose trials are classified w.r.t. a given family of properties $\mathbf{Q}=(Q_1,...,Q_k)$ - is a multivariate Bernoulli process with parameter vector $\mathbf{q}=(\mathbf{q}_1,...,\mathbf{q}_{k-1})$. In this case those researchers who wish to use the Bayesian approach should select a prior *cdf* $F(q_1,...,q_k)$ on $\mathbf{q}$ on the basis of *BK*.

With reference to this, recall that *BK* may include some external information $EIn(\mathbf{q},Ex)$ about $\mathbf{q}$ in *Ex*. In particular, *BK* may include (i) some *experimental information* on the value of $\mathbf{q}$ in processes 'sufficiently similar' to *Ex* (such as the statement "the value of $\mathbf{q}$ in the multivariate Bernoulli process *Ex'* is *a*"), or (ii) some *theoretical information* on the relations between the value of $\mathbf{q}$ and the values of other quantities $\boldsymbol{\phi}_1,...,\boldsymbol{\phi}_m$ in processes 'sufficiently similar' to *Ex* (together with information available on the values of $\boldsymbol{\phi}_1,...,\boldsymbol{\phi}_m$ in *Ex*).

Although a more precise description of the above external information would require specification of the concept of "sufficiently similar processes", an informal use of this concept is sufficient for our purposes. For instance, the multivariate Bernoulli processes consisting in throwing several dice of the same shape and made of the same material might be considered sufficiently similar to each other also in the case where the dice are of different size. From now on, a multivariate Bernoulli process *Ex'* will be referred to as an "*Ex-similar process*" in the case where it is sufficiently similar to *Ex*.

Below I will describe a situation where the external information $EIn(\mathbf{q},Ex)$ justifies the use of the requirement of adequacy (CM) and, consequently, the Dirichlet family.

Let us suppose that the experimental information contained in $EIn(\mathbf{q},Ex)$ includes the statements $H_1,...,H_m$ which specify the values of $\mathbf{q}$ for the *Ex*-similar processes $Ex_1,...,Ex_m$. Given a statement H_t, where $t = 1,...,m$, the value of $\mathbf{q}$ specified by H_t will be referred to as $a^t=(a^t_1,...,a^t_k)$, where a^t_i denotes the objective probability $\mathbf{q}_i$ of Q_i in Ex_t. Then the k - 1 fractions $a^t_j/(1 - a^t_i)$ (where $j \neq i$) specify how the 'residual part' $(1 - a^t_i)$ of $\mathbf{q}$ in Ex_t is divided among the remaining properties Q_j.

Given two distinct properties - Q_i and Q_j - let us consider (1) the quantities $a^1_i,...,a^m_i$ denoting the objective probabilities of Q_i in $Ex_1,...,Ex_m$ specified by the hypotheses $H_1,...,H_m$, respectively, and (2) the quantities $a^1_j/(1 - a^1_i),...,$ $a^m_j/(1 - a^m_i)$ expressing the ratios between the objective probabilities $a^1_j,...,a^m_j$ of Q_j in $Ex_1,...,Ex_m$ and the 'residual parts' $(1 - a^1_i),...,(1 - a^m_i)$ of $\mathbf{q}$ in $Ex_1,...,Ex_m$.

Now, each of the two sets $a^1_i,...,a^m_i$ and $a^1_j/(1 - a^1_i),...,a^m_j/(1 - a^m_i)$ will typically include values somewhat different from each other. However, suppose that the changes in the values of the ratios $a^t_j/(1 - a^t_i)$ - where $t = 1,...,m$ - *are not* appreciably related (in some systematic way) to certain corresponding changes in the values of a^t_i.[29] In this case we may say that *Q_j is $H_1,...,H_m$-independent of Q_i*.[30]

When m is sufficiently high, the $H_1,...,H_m$-independence of Q_j and Q_i suggests that the 'causal mechanism' producing *Ex*-similar processes works in such a way that, as it were, the 'production' of a high (low) ratio $\mathbf{q_i}$ of Q_i-outcomes neither favours nor hampers the 'production' of a high (low) residual ratio $\mathbf{q_j}/(1 - \mathbf{q_i})$ of Q_j-outcomes. This hypothesis may be called the hypothesis of *independence of Q_j and Q_i (for Ex-similar processes)*.

Suppose that the independence of Q_j and Q_i for *Ex*-similar processes is accepted. The acceptance of this hypothesis could be 'formally represented' by the following formal constraint c_{ji} on the prior *cdf* $F(q_1,...,q_{k-1})$ on $\mathbf{q}=(\mathbf{q_1},...,\mathbf{q_{k-1}})$ in *Ex*: "$\mathbf{q_j}/(1 - \mathbf{q_i})$ (where $j \neq i$) is independent of $\mathbf{q_i}$".[31] Notice that the conjunction of the k - 1 formal constraints, or requirements of adequacy, c_{ji} (where $j \neq i$) is equivalent to the $(CM)_i$-neutrality of $\mathbf{q}$ (cf. Chapter 6.6).

Now, assume that, given the experimental information $H_1,...,H_m$ included in *EIn*($\mathbf{q}$,*Ex*), researchers have established the $H_1,...,H_m$-independence of Q_j and Q_i, for any couple $Q_i \neq Q_j$. If the above arguments are valid, then this assumption implies that researchers:

(a) should accept the corresponding hypotheses of independence of Q_j an Q_i for *Ex*-similar processes;
(b) should adopt the corresponding constraints c_{ji} on $\mathbf{q}$;
(c) should assume the $(CM)_i$-neutrality of $\mathbf{q}$, for $i = 1,...,k$.

This is tantamount to assuming that the prior distribution $F(q_1,...,q_{k-1})$ on $\mathbf{q}$ satisfies the requirement of adequacy (CM) or, equivalently, that $F(q_1,...,q_{k-1})$ is a Dirichlet distribution (cf. 6.6). In other words the above contextual justification of (CM) can be interpreted as a contextual justification of the decision to use a prior Dirichlet.

Moreover, given the equivalence between Dirichlet distributions and GC-systems (cf. 6.3), the above contextual justification of (CM) can be seen as a contextual justification of the decision to adopt, within a given empirical research, an inductive method (γ,λ) belonging to the class of GC-systems.[32]

Having decided to use a GC-system (γ,λ), the values of the prior vector γ and of the parameter λ must be selected. A 'contextual analysis' of these decisions will be provided in the next chapter.

CHAPTER 8

THE EPISTEMIC PROBLEM OF OPTIMALITY (EPO): A CONTEXTUAL APPROACH

The problem of determining the optimum Dirichlet distribution - or, equivalently, the optimum GC-system - for investigating a given multivariate Bernoulli process *Ex* with parameter vector $\mathbf{q}=(q_1,...,q_k)$ may be referred to as *the epistemic problem of optimality*, or *EPO*. In this chapter a *contextual approach* to EPO is proposed.

Given the equivalence between GC-systems and Dirichlet distributions (see Chapter 6.3), only Dirichlet distributions will be dealt with. A Dirichlet distribution $\mathrm{Dir}(\alpha_1,...,\alpha_k)$ on $\mathbf{q}$ will be referred to as $\mathrm{Dir}(\gamma,\lambda)$, where $\gamma=(\gamma_1,...,\gamma_k) = (\alpha_1/\lambda,...,\alpha_k/\lambda)$ is the prior vector of $\mathrm{Dir}(\alpha_1,...,\alpha_k)$ and $\Sigma\alpha_i$ is denoted by λ, to recall visually that $\mathrm{Dir}(\alpha_1,...,\alpha_k) = \mathrm{Dir}(\gamma,\lambda)$ is equivalent to the GC-system (γ,λ).

Solving EPO for a given empirical research situation consists in establishing the optimum prior distribution $\mathrm{Dir}(\gamma,\lambda)$, i.e., the optimum values of γ and λ.

The selection of the optimum prior vector $\gamma°=(\gamma_1^\circ,....,\gamma_k^\circ)$ will be considered in Section 1.

Subsequent sections will deal with the problem of determining the optimum prior (and the related optimum λ-value) within the set $\mathrm{DIR}(\gamma°)$ of Dirichlet distributions $\mathrm{Dir}(\gamma°,\lambda)$ with prior vector $\gamma°$. Since selection of the optimum λ-value represents the most difficult part of EPO, this problem will hence be referred to as EPO.

In Section 2, an important type of multinomial cognitive context called *GD-context* is defined.

Later two 'contextual solutions' to EPO applicable to GD-contexts are proposed.

The first solution, illustrated in Section 3, is called a *cognitive coherence solution*, or *CC-solution*, since it is based on the requirement of cognitive coherence (CC).

The second solution, described in Sections 4-5, is called a *verisimilitude solution*, or *V-solution*, since it is based on the verisimilitude thesis (VER) which states that the most important cognitive goal of pure science is the achievement of a high degree of verisimilitude (see Chapter 1.1). The V-solution is obtained using the following strategy:

(a) Having defined the *logical problem of optimality*, or *LPO*, for the set DIR(γ) of all Dirichlet distributions with a given prior vector γ, a V-solution to LPO is provided (Section 4).

(b) Using (VER) and the V-solution to LPO, a V-solution to EPO is elaborated (Section 5).

Somewhat surprisingly, in spite of the conceptual differences between the V- and the CC-solution to EPO, they give identical results. It will be argued that this equivalence, far from being a fortunate mathematical coincidence, can be conceptually explained (Section 6).

Lastly, in Section 7 the relationships between the above-mentioned V-solution to LPO and some results obtained by Carnap (1952) are considered.

8.1 THE OPTIMUM PRIOR VECTOR

In several cases the optimum prior vector $\gamma°$ can be derived from the available background knowledge *BK*. For instance, suppose that the external information $EIn(\mathbf{q},Ex)$ in *BK* is given by the experimental information $H_1,...,H_m$, where such statements specify the values of $\mathbf{q}=(\mathbf{q}_1,...,\mathbf{q}_k)$ for the *Ex*-similar process $Ex_1,...,Ex_m$ (cf. Chapter 7.6). In particular, remember that the values of $\mathbf{q}_i$ specified by $H_1,...,H_m$ are referred to as $a_i^1,...,a_i^m$, respectively.

If m is not too low the following external estimate $Est(\mathbf{q}_i)$ of $\mathbf{q}_i$ in *Ex* may be made:

$$(1) \qquad Est(\mathbf{q}_i) = (a_i^1 + \cdots + a_i^m)/m = \frac{\sum_{t=1}^{m} a_i^t}{m}$$

Alternatively the arithmetical average $\Sigma a_i^t/m$ in (1) could be replaced by the following more sophisticated weighted average:

$$(2) \qquad Est(\mathbf{q}_i) = \frac{\sum_{t=1}^{m} w_t a_i^t}{W} \qquad (W = \sum_{t=1}^{m} w_t)$$

where the weights $w_1,...,w_m$ may reflect the varying 'degrees of similarity' between Ex and each of the Ex-similar process $Ex_1,...,Ex_m$ (and, possibly, the researchers' commitment to the 'accuracy' of each of the statements $H_1,...,H_m$).[1]

In the cases described above and, more generally, whenever the external estimates $Est(\mathbf{q}_1),...,Est(\mathbf{q}_k)$ are available, it seems reasonable to equate the marginal means of Dir(γ,λ) to the corresponding external estimates: $E(\mathbf{q}_1) = Est(\mathbf{q}_1),..., E(\mathbf{q}_k) = Est(\mathbf{q}_k)$.[2] Given the equalities $E(\mathbf{q}_i) = \gamma_i$ (see formula (3.32)) this is tantamount to taking $Est(\mathbf{q}) = (Est(\mathbf{q}_1),..., Est(\mathbf{q}_k))$ as the optimum prior vector $\gamma°=(\gamma_1°,...,\gamma_k°)$.

8.2 GD-CONTEXTS

As stated at the beginning of this chapter, the analysis of EPO made in Sections 3-5 will be made w.r.t. to a particular type of multinomial context called GD-contexts. Before defining GD-contexts, the notions of G-context and D-context must be introduced.

A *G-context* is a multinomial context where, in addition to an external estimate $Est(\mathbf{q})$ of $\mathbf{q}$ in Ex, an external estimate $Est[G(\mathbf{q})]$ of the Gini diversity $G(\mathbf{q}) = 1 - \Sigma \mathbf{q}_i^2$ can be made.

Consider, for instance, a multinomial context where the external information $EIn(\mathbf{q},Ex)$ in BK is given by the statements $H_1,...,H_m$ which specify the values of $\mathbf{q}$ for the Ex-similar process $Ex_1,...,Ex_m$ (cf. Section 8.1). Given the value $a^t=(a_1^t,...,a_k^t)$ of $\mathbf{q}$ specified by H_t, where $t = 1,...,m$, the corresponding Gini diversity $1 - \Sigma(a_i^t)^2$ will be referred to as $G(a^t)$. If m is not too low, the following external estimate $Est[G(\mathbf{q})]$ of $G(\mathbf{q})$ in Ex may be made:

$$(3) \qquad Est[G(\mathbf{q})] = \frac{\sum_{t=1}^{m} G(a^t)}{m}$$

Alternatively, the following weighted average could be used:

$$(4) \qquad Est[G(\mathbf{q})] = \frac{\sum_{t=1}^{m} w_t G(a^t)}{W}$$

where $w_1,\ldots,w_m$ and W are interpreted as in (2).[3]

In other G-contexts the external estimates $Est[G(\mathbf{q})]$ could be made using also the theoretical information contained in $EIn(\mathbf{q},Ex)$ (cf. Chapter 7.5). For instance, in ecology, an estimate $Est[G(\mathbf{q})]$ of the Gini diversity of a given population could be made using appropriate hypotheses (cf. Chapter 10.3.) about the relations between $G(\mathbf{q})$ and other characteristics of the population (such as stability) or its environment (such as pollution).

A *D-context* is a multinomial context where, in addition to $Est(\mathbf{q})$, an external estimate $Est[D_2(\gamma°,\mathbf{q})]$ of the distance $D_2(\gamma°,\mathbf{q})$ between the prior vector $\gamma° = Est(\mathbf{q})$ and the true value of $\mathbf{q}$ in Ex can be made.

Consider, for instance, the case where the available external information $EIn(\mathbf{q},Ex)$ is given by the above-mentioned statements $H_1,\ldots,H_m$ concerning the Ex-similar process $Ex_1,\ldots,Ex_m$. Here, if m is not too low, the following external estimate $Est[D_2(\gamma°,\mathbf{q})]$ could be made:

$$(5) \quad Est[D_2(\gamma°,\mathbf{q})] = \frac{\sum_{t=1}^{m} D_2(\gamma°,a^t)}{m}$$

Alternatively, the following weighted average could be used:

$$(6) \quad Est[D_2(\gamma°,\mathbf{q})] = \frac{\sum_{t=1}^{m} w_t D_2(\gamma°,a^t)}{W}$$

where $w_1,\ldots,w_m$ and W are interpreted as in (2) and (4).[4]

Now a *GD-context* can be defined as a multinomial context which is both a G- and a D-context. In other words a GD-context is a multinomial context where, in addition to $Est(\mathbf{q})$, both the external estimates $Est[G(\mathbf{q})]$ and $Est[D_2(\gamma°,\mathbf{q})]$ can be made. Since $Est(\mathbf{q})$, $Est[G(\mathbf{q})]$ and $Est[D_2(\gamma°,\mathbf{q})]$ are all that is needed to choose the optimum prior vector $\gamma°$ (see Section 1) and the optimum λ-value $\lambda°$ (see Sections 3-5), it follows that the background knowledge BK available in a GD-context is sufficient to identify the optimum GC-system $(\gamma°,\lambda°)$, i.e., to provide a complete solution to EPO.

In Chapter 7.6, it was seen that the empirical information conveyed by the statements $H_1,...,H_m$ can be used, if m is sufficiently high, to decide whether the inductive method to be employed within a given empirical research should or should not be a GC-system. From (1)-(6) it appears that $H_1,...,H_m$ can also be used to derive the external estimates $Est(\mathbf{q})$, $Est[G(\mathbf{q})]$, and $Est[D_2(\gamma°,\mathbf{q})]$ which in turn can be used to solve EPO, i.e. to specify the optimum GC-system $(\gamma°,\lambda°)$ (see Sections 3-5). This signifies that the same experimental information $H_1,...,H_m$ used to decide *whether* a GC-system should be adopted can be 'reprocessed' to establish *which* GC-system should be adopted.

It follows from the requirement of adequacy (Ext)(a) (see Chapter 7.6, p. 98) that $\gamma° = Est(\mathbf{q})$, $Est[G(\mathbf{q})]$ and $Est[D_2(\gamma°,\mathbf{q})]$ are linked by the following equality:

(7) $$Est[D_2(\gamma°,\mathbf{q})] = G(\gamma°) - Est[G(\mathbf{q})]$$

Proof of (7): Given the equalities $D_2(\gamma°,\mathbf{q}) = \Sigma(\gamma_i° - \mathbf{q}_i)^2$ and $\gamma_i° = Est(\mathbf{q}_i)$, it follows from (Ext)(a) that $Est[D_2(\gamma°,\mathbf{q})] = Est[\Sigma(\gamma_i°)^2 + \Sigma\mathbf{q}_i^2 - 2\Sigma\gamma_i°\mathbf{q}_i] = Est(\Sigma\mathbf{q}_i^2) + \Sigma(\gamma_i°)^2 - 2\Sigma\gamma_i° Est(\mathbf{q}_i) = Est(\Sigma\mathbf{q}_i^2) - \Sigma(\gamma_i°)^2 = [1 - \Sigma(\gamma_i°)^2] - [1 - Est(\Sigma\mathbf{q}_i^2)]$. From this sequence of equalities, and taking into account that $[1-\Sigma(\gamma_i°)^2] = G(\gamma°)$ and $[1 - Est(\Sigma\mathbf{q}_i^2)] = Est(1 - \Sigma\mathbf{q}_i^2) = Est[G(\mathbf{q})]$, it follows that $Est[D_2(\gamma°,\mathbf{q})] = G(\gamma°) - Est[G(\mathbf{q})]$, *Q.E.D.*[5]

In GD-contexts the external estimates $Est(\mathbf{q})$, $Est[G(\mathbf{q})]$, and $Est[D_2(\gamma°,\mathbf{q})]$ can be made using specific procedures such as those defined in (1)-(6). The equality (7) can be seen as a 'constraint', or requirement of adequacy, on such procedures. In particular, it can be proved that such a constraint is satisfied by the procedures defined in (1), (3), and (5).[6] Analogously, it is also satisfied by the 'weighted' external estimates suggested in (2), (4), and (6).[7]

On the other hand, in those G-contexts where no 'direct procedure' for the external estimate of $D_2(\gamma°,\mathbf{q})$ is available the equality (7) provides an 'indirect procedure' for deriving $Est[D_2(\gamma°,\mathbf{q})]$ (see remarks below (Ext), p. 98). This implies that any G-context is also a D-context and, consequently, a GD-context. Similarly, any D-context is also a GD-context.[8] Hence, the distinction between G-contexts, D-contexts and GD-contexts is only a *prima facie* distinction since all these contexts are, indeed, GD-contexts.[9]

8.3 THE CC-SOLUTION TO EPO

The CC-solution to EPO for GD-contexts is based on the requirement of cognitive coherence (CC).

As it was suggested in the last paragraph of Section 1, whenever the external estimates $Est(\mathbf{q}_1),\ldots,Est(\mathbf{q}_k)$ are available, the marginal means of $\text{Dir}(\gamma,\lambda)$ can be equated to the corresponding external estimates: $E(\mathbf{q}_1) = Est(\mathbf{q}_1),\ldots, E(\mathbf{q}_k) = Est(\mathbf{q}_k)$ where - given the equalities $E(\mathbf{q}_i) = \gamma_i$ - this is tantamount to taking $Est(\mathbf{q}) = (Est(\mathbf{q}_1),\ldots, Est(\mathbf{q}_k))$ as the optimum prior vector $\gamma^\circ=(\gamma_1^\circ,\ldots,\gamma_k^\circ)$. In note 1 it was pointed out that this procedure for the selection of the optimum prior vector γ° can be seen as an application of the requirement of cognitive coherence (CC).

One can show that in a GD-context - given a certain value of $\gamma^\circ = Est(\mathbf{q})$ - the requirement (CC) can be used also for solving EPO, i.e., for determining the optimum prior (and the related optimum λ-value) within the set $\text{DIR}(\gamma^\circ)$ of Dirichlet distributions $\text{Dir}(\gamma^\circ,\lambda)$ with prior vector γ°.

Indeed, given the external estimate $Est[G(\mathbf{q})]$ available in a certain GD-context, the requirement of cognitive coherence (CC) imposes the following m-constraint c_G on $\text{Dir}(\gamma^\circ,\lambda)$: c_G = "$E[G(\mathbf{q})] = Est[G(\mathbf{q})]$". Taking account of the equality $\lambda = E[G(\mathbf{q})]/(G(\gamma^\circ) - E[G(\mathbf{q})])$ - derived from formula (7.1) by replacing α and γ with λ and γ° - the constraint c_G implies that λ is given by:

$$(8) \qquad \lambda = \frac{Est[G(\mathbf{q})]}{G(\gamma^\circ) - Est[G(\mathbf{q})]}$$

Similarly, given the external estimate $Est[D_2(\gamma^\circ,\mathbf{q})]$ of $D_2(\gamma^\circ,\mathbf{q})$, (CC) imposes the following m-constraint c_D on $\text{Dir}(\gamma^\circ,\lambda)$: c_D = "$E[D_2(\gamma^\circ,\mathbf{q})] = Est[D_2(\gamma^\circ,\mathbf{q})]$". Taking account of the equality $\lambda = (G(\gamma^\circ) - E[D_2(\gamma^\circ,\mathbf{q})])/E[D_2(\gamma^\circ,\mathbf{q})]$ - derived from formula (7.2) by replacing α and γ with λ and γ° - the constraint c_D implies that λ is given by:

$$(9) \qquad \lambda = \frac{G(\gamma^\circ) - Est[D_2(\gamma^\circ,\mathbf{q})]}{Est[D_2(\gamma^\circ,\mathbf{q})]}$$

Note that the application of (CC) in GD-contexts cannot lead to the acceptance of two incompatible constraints c_G and c_D which 'overdetermine' the problem of identifying the value of λ. Indeed, it directly follows from equality (7) that formulae (8) and (9) should provide the same λ-value which will hence be referred to as *the CC-optimum λ-value* λ_{CC}. Note that the value of λ_{CC} can be also expressed by the following more simple formula:

$$(10) \qquad \lambda_{CC} = \frac{Est[G(\mathbf{q})]}{Est[D_2(\gamma^\circ,\mathbf{q})]}$$

Formula (9) shows that, given a certain value of $G(\gamma^\circ)$, when $Est[D_2(\gamma^\circ,\mathbf{q})]$ increases then the CC-optimum value λ_{CC} decreases.[10] The intuitive plausibility of this behaviour can be explained as follows.

Firstly, given a certain value of $G(\gamma^\circ)$, consider why, if $Est[D_2(\gamma^\circ,\mathbf{q})]$ is low, it is reasonable to select a high value of λ_{CC}. Recalling that $Est[D_2(\gamma^\circ,\mathbf{q})]$ can be seen as the estimate - based on the available background knowledge *BK* - of the distance from the truth, or 'inaccuracy', of γ° (cf. note 4), a low value of $Est[D_2(\gamma^\circ,\mathbf{q})]$ amounts to guessing that $\gamma^\circ = Est(\mathbf{q})$ is pretty accurate, i.e., very near to the true value of **q** (the truth). Now it appears perfectly natural to represent this guess by selecting - within the set DIR(γ°) of Dirichlet distributions Dir(γ°,λ) with prior vector γ° - a distribution Dir($\gamma^\circ,\lambda_{CC}$) with a high value of λ_{CC}. This prior Dirichlet distribution, indeed, attributes a very large amount of the whole probability to the region around the optimum prior vector γ°.

Secondly, given a certain value of $G(\gamma^\circ)$, consider why, if $Est[D_2(\gamma^\circ,\mathbf{q})]$ is high, it is reasonable to select a low value of λ_{CC}. According to the above mentioned interpretation of $Est[D_2(\gamma^\circ,\mathbf{q})]$ a high value of $Est[D_2(\gamma^\circ,\mathbf{q})]$ amounts to guessing that $\gamma^\circ = Est(\mathbf{q})$ is pretty inaccurate, i.e., quite far from the true value of **q** (the truth). Now it appears perfectly natural to represent this guess by selecting - within the set DIR(γ°) - a distribution Dir($\gamma^\circ,\lambda_{CC}$) with a low value of λ_{CC}. In fact this prior Dirichlet distribution does not attribute a very large amount of probability to the region around γ° but, on the contrary, spreads out the whole probability in a relative uniform manner over all the possible values of **q**.[11]

8.4 THE LOGICAL PROBLEM OF OPTIMALITY

Some general considerations

Suppose that, within an empirical research about a given parameter (vector) $\boldsymbol{\theta}$, a researcher *X* intends to

(i) perform an experiment **y**;

(ii) apply an estimator T for estimating $\boldsymbol{\theta}$ on the basis of the possible results y of $\mathbf{y}$ (see Chapter 4.1), and

(iii) use a distance function $D(T(\boldsymbol{\theta} \mid y),\theta)$ as a measure of his *cognitive failure* in the case where $\boldsymbol{\theta} = \theta$ and $\mathbf{y} = y$.[12]

Assume that $\boldsymbol{\theta}$ and $\mathbf{y}$ are linked by a probability function $P_\theta(y)$ which represents the *physical* (or objective) probability of the occurrence of y when $\mathbf{y}$ is performed in the case where $\boldsymbol{\theta} = \theta$. Then, given a value θ of $\boldsymbol{\theta}$, the expected (w.r.t. $P_\theta(y)$) distance $E(D(T(\boldsymbol{\theta} \mid \mathbf{y}),\theta) \mid \theta)$ is given by:

$$(11) \qquad E(D(T(\boldsymbol{\theta} \mid \mathbf{y}),\theta) \mid \theta) = \int P_\theta(y)\, D(T(\boldsymbol{\theta} \mid y),\theta)\, dy$$

The quantity $E(D(T(\boldsymbol{\theta} \mid \mathbf{y}),\theta) \mid \theta)$ can be seen as a measure of the *prospect of (cognitive) failure* of the estimator T in the state of nature θ relative to the experiment $\mathbf{y}$. Note that $E(D(T(\boldsymbol{\theta} \mid \mathbf{y}),\theta) \mid \theta)$ does not concern the beliefs of X - or of any other researcher - about $\boldsymbol{\theta}$, since it does not involve any epistemic probability but only the *physical* probability $P_\theta(y)$. Rather, $E(D(T(\boldsymbol{\theta} \mid \mathbf{y}),\theta) \mid \theta)$ may be seen as the description of a particular aspect of the 'structure of the world', i.e., the *actual*, or *objective*, prospect of failure of the estimator T in the state of nature θ, relative to the experiment $\mathbf{y}$. In other words, $E(D(T(\boldsymbol{\theta} \mid \mathbf{y}),\theta) \mid \theta)$ is not an epistemic notion but an *ontic notion* concerning, so to speak, an 'objective feature' of the triple T, θ, $\mathbf{y}$.

Given a triple $(\mathbf{T},\theta,\mathbf{y})$, where $\mathbf{T}$ is a given class of estimators, $T^\Delta \in \mathbf{T}$ can be defined as the *optimum estimator* relative to $(\mathbf{T},\theta,\mathbf{y})$ iff it minimizes the prospect of failure $E(D(T(\boldsymbol{\theta} \mid \mathbf{y}),\theta) \mid \theta)$. Analogous to $E(D(T(\boldsymbol{\theta} \mid \mathbf{y}),\theta) \mid \theta)$, the notion of optimum estimator T^Δ is a non-epistemic notion referring to an 'objective feature' of $(\mathbf{T},\theta,\mathbf{y})$. Such a notion could be seen as a *logical notion* because it is defined exclusively w.r.t. the 'logical relations' among $\mathbf{T}$, θ, and $\mathbf{y}$. Accordingly, the problem of identifying the optimum estimator T^Δ for a given triple $(\mathbf{T},\theta,\mathbf{y})$ will be called *the logical problem of optimality*, or *LPO*, for $(\mathbf{T},\theta,\mathbf{y})$.

The intuitive meaning of LPO may be elucidated from the following 'quasi-theological' situation. Imagine that:

(i) Researcher X, in his empirical inquiry on $\boldsymbol{\theta}$, intends to perform an experiment $\mathbf{y}$ and to use an estimator selected from a given class $\mathbf{T}$.

(ii) You are a semi-omniscient demigod in the sense that You know the true value θ of $\boldsymbol{\theta}$ and, although not knowing which particular value of y will be observed by X, You do know the (true) physical probability $P_\theta(y)$.

(iii) You know the distance function $D(T(\mathbf{\theta} \mid y),\theta)$ used by X as a measure of cognitive failure and, consequently, You know the prospect of cognitive failure $E(D(T(\mathbf{\theta} \mid \mathbf{y}),\theta) \mid \theta)$ of any estimator $T \in \mathbf{T}$ in the actual state of nature θ.

(iv) Being benevolent towards X - as a demigod should be - You want to help him in pursuing his cognitive goals. Suppose that, for some reason, You cannot reveal to X the true value of $\mathbf{\theta}$ but You can only suggest to him to use a given estimator from $\mathbf{T}$. Which one would You suggest? It seems clear that You would suggest to X the estimator T^{Δ} which minimizes the prospect of failure $E(D(T(\mathbf{\theta} \mid \mathbf{y}),\theta) \mid \theta)$ in the actual state of nature.

(v) Accordingly, the following 'quasi-theological' interpretation of the logical notion of optimum estimator could be proposed: the optimum estimator T^{Δ} relative to $(\mathbf{T},\theta,\mathbf{y})$, is that which would be suggested by a semi-omniscient and benevolent demigod to a researcher who intends to use the result of the experiment $\mathbf{y}$ for estimating $\mathbf{\theta}$.

A solution to LPO for certain triples $(\mathbf{T},\theta,\mathbf{y})$ may be *theoretically* interesting. But what of its *practical* relevance? In other words: may a solution to LPO help to solve the 'corresponding' EPO, i.e., the problem of selecting, among the estimators in $\mathbf{T}$, the '*epistemically* optimum' for the empirical inquiry on $\mathbf{\theta}$ made by X?

Of course, if X knew the logically optimum estimator for the actual state of nature, he would select it as the epistemically optimum estimator for his inquiry on $\mathbf{\theta}$. However, in the absence of a benevolent and semi-omniscient demigod, X cannot identify the logically optimum estimator w.r.t. the *actual* state of nature but can only establish, by solving LPO, the logically optimum estimator for any value θ of $\mathbf{\theta}$, i.e., for any *possible* state of nature.

Although solving LPO does not directly offer a solution to EPO, in certain multinomial contexts it is a crucial element in the formulation of an adequate solution to EPO (see Section 5).

Logically optimum Dirichlet distributions

Consider the prior distribution $F(q_1,...,q_{k-1})$ on the parameter vector $\mathbf{q}=(\mathbf{q}_1,...,\mathbf{q}_k)$ of a multivariate Bernoulli process Ex and the experiment $\mathbf{e}_n$ which consists in giving a complete description of a sequence of n trials of Ex, where e_n denotes a possible result of $\mathbf{e}_n$. Moreover, let E be the

$F(q_1,...,q_{k-1})$-based M-estimator which takes, for any e_n, the mean vector $E(\mathbf{q} \mid e_n)=(E(\mathbf{q}_1/e_n),..., E(\mathbf{q}_k/e_n))$ of the posterior distribution $F(q_1,...,q_{k-1}/e_n)$ as the point estimate of $\mathbf{q}$ (see Chapter 4.2). Here it is assumed that $F(q_1,...,q_{k-1}/e_n)$ is derived from the prior distribution $F(q_1,...,q_{k-1})$ and the likelihood function $p(e_n/q) = P_q(e_n)$ where $P_q(e_n) = \Pi\, q_i^{n_i}$ represents the *physical* (or objective) probability of e_n (see formula (2.2)).

Then, given a prior distribution Dir(γ,λ) on $\mathbf{q}$, the Dir(γ,λ)-based M-estimator will be referred to as E_λ and the Dir(γ,λ)-based M-estimates of $\mathbf{q}$ as $E_\lambda(\mathbf{q} \mid e_n)=(E_\lambda(\mathbf{q}_1/e_n),..., E_\lambda(\mathbf{q}_k/e_n))$.

Now suppose that X, in an attempt to estimate the value of $\mathbf{q}$, intends to

(i) perform experiment $\mathbf{e}_n$;

(ii) apply the Dir(γ,λ)-based M-estimator E_λ for estimating $\mathbf{q}$ on the basis of the possible results e_n of $\mathbf{e}_n$, and

(iii) use the quadratic distance $D_2(E_\lambda(\mathbf{q} \mid e_n),q) = \Sigma[E_\lambda(\mathbf{q}_i/e_n) - q_i]^2$ as a measure of his *cognitive failure* in the case where $\mathbf{q} = q$ and $\mathbf{e}_n = e_n$.[13]

Given a value q of $\mathbf{q}$, the expected (w.r.t. $P_q(e_n)$) quadratic distance $E(D_2(E_\lambda(\mathbf{q} \mid \mathbf{e}_n),q) \mid q)$ can be seen as a measure of the *prospect of (cognitive) failure* of the M-estimator E_λ in the state of nature q relative to the experiment $\mathbf{e}_n$.

Now it follows from (11) and (3.16.b) that $E(D_2(E_\lambda(\mathbf{q} \mid \mathbf{e}_n),q) \mid q)$ is given by:

$$(12)\quad E(D_2(E_\lambda(\mathbf{q} \mid \mathbf{e}_n),q) \mid q) = E[\sum_{i=1}^{k} (E_\lambda(\mathbf{q}_i \mid \mathbf{e}_n) - q_i)^2 \mid q]$$

$$= \sum_{i=1}^{k} E[(E_\lambda(\mathbf{q}_i \mid \mathbf{e}_n) - q_i)^2 \mid q]$$

$$= \sum_{i=1}^{k} (\Sigma\, P_q(e_n)\, (E_\lambda(\mathbf{q}_i \mid \mathbf{e}_n) - q_i)^2)$$

where the inner sum is the expected (w.r.t $P_q(e_n)$) value of $(E_\lambda(\mathbf{q}_i \mid \mathbf{e}_n) - q_i)^2$ for a given value q of $\mathbf{q}$. Note that the inner summation extends over all possible results e_n of $\mathbf{e}_n$.

The quantities $D_2(E_\lambda(\mathbf{q} \mid \mathbf{e}_n),q)$, $(E_\lambda(\mathbf{q}_i \mid \mathbf{e}_n) - q_i)^2$, $E_\lambda(\mathbf{q}_i \mid \mathbf{e}_n) - q_i$ and $(E_\lambda(\mathbf{q}_i \mid \mathbf{e}_n)$ occurring in (12) are $\mathbf{e}_n$-related *statistics*. In particular, $E_\lambda(\mathbf{q}_i \mid \mathbf{e}_n)$ -

q_i and $(E_\lambda(\mathbf{q_i} \mid \mathbf{e_n}) - q_i)^2$ are called the *error* and the *square error*, respectively, of the *point estimate* $E_\lambda(\mathbf{q_i} \mid \mathbf{e_n})$ relative to the value q_i of $\mathbf{q_i}$.

The sampling distributions of the above statistics can be calculated from the probability distribution $P_q(e_n)$ (see (2.2)). In particular, it follows from (3.20.b) and (3.20.c) that the expected value $E[(E_\lambda(\mathbf{q_i} \mid \mathbf{e_n}) - q_i)^2 \mid q]$ is:

$$(13) \quad E[(E_\lambda(\mathbf{q_i} \mid \mathbf{e_n}) - q_i)^2 \mid q]$$

$$= \mathrm{var}(E_\lambda(\mathbf{q_i} \mid \mathbf{e_n}) \mid q) + [E(E_\lambda(\mathbf{q_i} \mid \mathbf{e_n}) - q_i) \mid q]^2$$

To calculate $[E(E_\lambda(\mathbf{q_i} \mid \mathbf{e_n}) - q_i)) \mid q]^2$ remember that $E_\lambda(\mathbf{q_i} \mid e_n)$ is a linear function of n_i (cf. (6.4)):

$$(14) \quad E_\lambda(\mathbf{q_i} \mid e_n) = \frac{1}{n+\lambda} n_i + \frac{\lambda \gamma_i}{n+\lambda}$$

In fact it follows from (14), (3.16.a) and (3.24), that the mean $E[E_\lambda(\mathbf{q_i} \mid \mathbf{e_n}) \mid q]$ of $E_\lambda(\mathbf{q_i} \mid \mathbf{e_n})$ is given by:

$$(15) \quad E[E_\lambda(\mathbf{q_i} \mid \mathbf{e_n}) \mid q] = \frac{1}{n+\lambda} E_q(\mathbf{n_i}) + \frac{\lambda \gamma_i}{n+\lambda} = \frac{nq_i + \lambda \gamma_i}{n+\lambda}$$

Given (15) and (3.16.a), the following formula for the expected value $E(E_\lambda(\mathbf{q_i} \mid \mathbf{e_n}) - q_i \mid q)$ of $E_\lambda(\mathbf{q_i} \mid \mathbf{e_n}) - q_i$ can be obtained:

$$(16) \quad E(E_\lambda(\mathbf{q_i} \mid \mathbf{e_n}) - q_i \mid q) = \frac{\lambda(\gamma_i - q_i)}{n+\lambda}$$

Moreover, it follows from (14), (3.20.c) and (3.25) that the variance $\mathrm{var}(E_\lambda(\mathbf{q_i} \mid \mathbf{e_n}) \mid q)$ of $E_\lambda(\mathbf{q_i} \mid \mathbf{e_n}) - q_i$ is given by:

$$(17) \quad \mathrm{var}(E_\lambda(\mathbf{q_i} \mid \mathbf{e_n}) \mid q) = \mathrm{var}\left[\frac{1}{n+\lambda} \mathbf{n_i} + \frac{\lambda \gamma_i}{n+\lambda}\right]$$

$$= \frac{1}{(n+\lambda)^2} \mathrm{var}_q(\mathbf{n_i}) = \frac{nq_i(1-q_i)}{(n+\lambda)^2}$$

Given (13), (16) and (17), the following formula can be obtained:

$$(18) \qquad E[(E_\lambda(\mathbf{q_i} \mid \mathbf{e_n}) - q_i)^2 \mid q] = \frac{nq_i(1 - q_i)}{(n + \lambda)^2} + \frac{\lambda^2(\gamma_i - q_i)^2}{(n + \lambda)^2}$$

Finally, it can be deduced from (12) and (18), after a number of routine calculations, that:

$$(19) \qquad E(D_2(E_\lambda(\mathbf{q} \mid \mathbf{e_n}),q) \mid q) = \frac{n(1 - \Sigma\, q_i^2) + \lambda^2\Sigma\,(\gamma_i - q_i)^2}{(\text{n} + \lambda)^2}$$

Recalling that $G(q) = (1 - \Sigma q_i^2)$ (see Chapter 3.4) and that $D_2(\gamma,q) = \Sigma(\gamma_i - q_i)^2$ (see Chapter 4.1) formula (19) can be rewritten thus:

$$(20) \qquad E(D_2(E_\lambda(\mathbf{q} \mid \mathbf{e_n}),q) \mid q) = \frac{nG(q) + \lambda^2 D_2(\gamma,q)}{(\text{n} + \lambda)^2}$$

Formula (20) reveals two positive aspects of $E(D_2(E_\lambda(\mathbf{q} \mid \mathbf{e_n}),q) \mid q)$. Firstly, as would be natural to expect, the prospect of cognitive failure $E(D_2(E_\lambda(\mathbf{q} \mid \mathbf{e_n}),q) \mid q)$ of the *M*-estimator E_λ in the state of nature q, relative to the experiment $\mathbf{e_n}$, depends on n, λ, the prior vector $\gamma=(\gamma_1,...,\gamma_k)$ and the 'objective vector' $q=(q_1,...,q_k)$. Secondly, $E(D_2(E_\lambda(\mathbf{q} \mid \mathbf{e_n}),q) \mid q)$ depends on γ and q only via the Gini diversity $G(q)$ and the quadratic distance $D_2(\gamma,q)$.

Consider the set DIR(γ) of the Dirichlet distributions Dir(γ,λ) with prior vector γ, and the corresponding set $\mathbf{E}_\gamma$ containing the Dir(γ,λ)-based *M*-estimators E_λ. To solve LPO for a given triple $(\mathbf{E}_\gamma,q,\mathbf{e_n})$ the *M*-estimator which minimizes the prospect of failure $E(D_2(E_\lambda(\mathbf{q} \mid \mathbf{e_n}),q) \mid q)$ in $\mathbf{E}_\gamma$ must be identified. In other words, the solution to LPO for $(\mathbf{E}_\gamma,\text{q},\mathbf{e_n})$ is the *M*-estimator E^Δ with the λ-value λ^Δ which minimizes $[nG(q) + \lambda^2 D_2(\gamma,q)]/(n + \lambda)^2$, i.e., the value of λ for which the λ-derivative of this expression is 0. This leads to the conclusion that the *logically optimum* λ*-value* λ^Δ is given by:

$$(21) \qquad \lambda^\Delta = \frac{G(q)}{D_2(\gamma,q)}$$

The prior distribution Dir(γ,λ^Δ) will be called the *logically optimum Dirichlet* within DIR(γ).

Formula (21) reveals a positive feature of λ^{Δ}, i.e., the optimum value λ^{Δ} *does not* depend on n even though the prospect of failure $E(D_2(E_\lambda(\mathbf{q} \mid \mathbf{e}_n),q) \mid q)$ of any M-estimator E_λ *does* depend on n (cf. (20)). Therefore the λ-value λ^{Δ} optimum for a given experiment $\mathbf{e}_n$ is optimum *tout court*.

Formula (21) also shows that λ^{Δ} is directly proportional to $G(q)$ and inversely proportional to $D_2(\gamma,q)$. An 'intuitive interpretation' of this behaviour is given below.

Firstly, given a certain value of $G(q)$, consider why if $D_2(\gamma,q)$ is low it is reasonable to expect λ^{Δ} to be high. Remember that if, and only if, λ is high, $E_\lambda(\mathbf{q}_i \mid e_n)$ tends to have a value close to γ_i (cf. (6.4)). This signifies that an M-estimator E_λ with a high λ-value is 'rigid' in the sense that the point estimates $E_\lambda(\mathbf{q} \mid e_n)$ tend to have values close to the prior vector γ. It is clear that if γ and q are near to each other a rigid M-estimator helps to make estimates approaching q. In other words, if there is a small distance $D_2(\gamma,q)$ between γ and q, then only a high λ-value can be a 'good instrument' for approaching the truth (the true value q of $\mathbf{q}$). Hence, only a high λ-value can be the 'optimum instrument' λ^{Δ} for approaching the truth.

Secondly, given a certain value of $D_2(\gamma,q)$, consider why if $G(q)$ is low it is reasonable to expect λ^{Δ} to be low. Remember that if, and only if, λ is low, $E_\lambda(\mathbf{q}_i \mid e_n)$ tends to approach n_i/n (cf. (6.4)). This signifies that an M-estimator E_λ with a low λ-value is 'flexible' in the sense that the point estimates $E_\lambda(\mathbf{q}/e_n)$ tend to approach the *empirical vector* $v(e_n)=(n_1/n,\dots,n_k/n)$. It is clear that if there is a high physical probability that the empirical vectors $v(e_n)$ related to the experimental results e_n of $\mathbf{e}_n$ are near to q, then a flexible M-estimator helps to make estimates approaching q. More precisely, if the expected (w.r.t. $P_q(e_n)$) value $E(D_2(v(\mathbf{e}_n),q) \mid q)$ of the distance $D_2(v(\mathbf{e}_n),q)$ is low, then only a low λ-value can be a 'good instrument' for approaching the truth. Now it can be proved that $E(D_2(v(\mathbf{e}_n),q) \mid q)$ is directly proportional to $G(q)$:[14]

$$(22)\qquad E(D_2(v(\mathbf{e}_n),q) \mid q) = \frac{G(q)}{n}$$

This implies that, if there is a small Gini diversity $G(q)$, then only a low λ-value can be a 'good instrument' for approaching the truth (the true value q of $\mathbf{q}$). Hence, only a low λ-value can be the 'optimum instrument' λ^{Δ} for approaching the truth.

8.5 THE V-SOLUTION TO EPO

Suppose that the cognitive context **C** of researcher X investigating **q** is a GD-context. Solving EPO for **C** is equal to determining the optimum prior distribution within the set DIR(γ°) of Dirichlet distributions Dir(γ°,λ) with prior vector γ° (= $Est(\mathbf{q})$).

A verisimilitude solution, or V-solution, to EPO for **C** may be based on a suitable combination of the above V-solution to LPO and the verisimilitude thesis (VER) which takes the achievement of a high degree of verisimilitude as the aim of pure science (see Chapter 1.1). Indeed it is a basic assumption of the V-solution to EPO that X should use the external estimates $Est[G(\mathbf{q})]$ and $Est[D_2(\gamma,\mathbf{q})]$ available in **C** in an attempt to approach the truth.

In particular, the following two step V-solution to EPO will be referred to as V_1:

(i) X assumes that $G(q) = Est[G(\mathbf{q})]$ and $D_2(\gamma^\circ,q) = Est[D_2(\gamma^\circ,\mathbf{q})]$ and, accordingly,

(ii) X takes the distribution in DIR(γ°), which would be logically optimum *if* the above assumptions were true, as the epistemically optimum prior.

The epistemically optimum prior suggested by V_1 will henceforth be referred to as Dir($\gamma^\circ,\lambda_1^\circ$) where λ_1° is *the* V_1*-optimum* λ*-value*. The value of λ_1° can be calculated as follows.

Given (20), the prospect of cognitive failure $E(D_2(E_\lambda(\mathbf{q} \mid \mathbf{e}_n),q) \mid q)$ of a Dir(γ°,λ)-based M-estimator in q depends exclusively on $G(q)$ and $D_2(\gamma^\circ,q)$ and in particular:

(23) The value of $E(D_2(E_\lambda(\mathbf{q} \mid \mathbf{e}_n),q) \mid q)$ in a 'possible world' q where $G(q) = Est[G(\mathbf{q})]$ and $D_2(\gamma^\circ,q) = Est[D_2(\gamma^\circ,\mathbf{q})]$ is given by the formula:

$$E(D_2(E_\lambda(\mathbf{q} \mid \mathbf{e}_n),q) \mid q) = \frac{nEst[G(\mathbf{q})] + \lambda^2 Est[D_2(\gamma^\circ,\mathbf{q})]}{(n + \lambda)^2}$$

From formulae (20) and (21) it follows that the λ-value λ_1° which minimizes $E(D_2(E_\lambda(\mathbf{q} \mid \mathbf{e}_n),q) \mid q)$ in (23) - i.e., the logically optimum value of λ for a value of q where $G(q) = Est[G(\mathbf{q})]$ and $D_2(\gamma^\circ,q) = Est[D_2(\gamma^\circ,\mathbf{q})]$ - is given by:

$$(24) \qquad \lambda_1^\circ = \frac{Est[G(\mathbf{q})]}{Est[D_2(\gamma^\circ,\mathbf{q})]}$$

Formula (24) reveals that the epistemically optimum value λ_1° is independent of n. Therefore X's choice of the optimum prior distribution is independent of the length n of the experiment $\mathbf{e}_n$ to be performed.

Formulae (24) and (14) illustrate that, somewhat surprisingly, the V_1-optimum λ-value λ_1° is identical to the CC-optimum λ-value λ_{CC}. While at this stage of the analysis the identity $\lambda_1^\circ = \lambda_{CC}$ hardly appears more than a lucky coincidence, the reasons for this will be explained later (see Section 6).

It could be argued that V_1 neglects the inevitably conjectural nature of any estimate. In fact step (i) of V_1 seems to suggest that X is certain that the external estimates $Est[G(\mathbf{q})]$ and $Est[D_2(\gamma^\circ,\mathbf{q})]$ are correct. However, this conflicts with the generally held idea that any inductive estimate is no more than reasonable conjecture.

Even though V_1 may be considered as somewhat 'naive', the V-approach to EPO need not be dismissed. Indeed a more 'sophisticated' V-solution to EPO - henceforth referred to as V_2 - can be obtained.

For this purpose remember that, given an arbitrary distribution $\mathrm{Dir}(\gamma^\circ,\lambda) \in \mathrm{DIR}(\gamma^\circ)$, the prospect of failure of the $\mathrm{Dir}(\gamma^\circ,\lambda)$-based M-estimator E_λ in a possible state of nature q (a possible value q of $\mathbf{q}$), relative to a certain experiment $\mathbf{e}_n$, is referred to as $E(D_2(E_\lambda(\mathbf{q} \mid \mathbf{e}_n),q) \mid q)$. Accordingly, an *external estimate of the prospect of failure* of E_λ in the unknown actual state of nature (the true value of $\mathbf{q}$) may be referred to as $Est[E(D_2(E_\lambda(\mathbf{q} \mid \mathbf{e}_n),\mathbf{q})]$.

As in V_1, in V_2 it is assumed that a researcher X should use the external estimates $Est[G(\mathbf{q})]$ and $Est[D_2(\gamma,\mathbf{q})]$ available in the GD-context $\mathbf{C}$ of his empirical research on $\mathbf{q}$ in his attempt to approach the truth. More specifically, according to the V-solution V_2 to EPO the following two step procedure should be adopted:

(i) X uses $Est[G(\mathbf{q})]$ and $Est[D_2(\gamma^\circ,\mathbf{q})$ to derive an external estimate $Est[E(D_2(E_\lambda(\mathbf{q} \mid \mathbf{e}_n),\mathbf{q})]$ for any distribution $\mathrm{Dir}(\gamma^\circ,\lambda) \in \mathrm{DIR}(\gamma^\circ)$,
(ii) X then takes the distribution in $\mathrm{DIR}(\gamma^\circ)$ which minimizes the above mentioned external estimate as the epistemically optimum prior.

Contrary to how it appears, step (i) is not particularly problematical since it follows from (20) and (Ext)(a) (Chapter 7.6, p. 98) that the value of $Est[E(D_2(E_\lambda(\mathbf{e}_n),\mathbf{q})]$ is given by:

$$(25)\qquad Est[E(D_2(E_\lambda(\mathbf{q} \mid \mathbf{e}_n),\mathbf{q})] = \frac{nEst[G(\mathbf{q})] + \lambda^2 Est[D_2(\gamma^\circ,\mathbf{q})]}{(n+\lambda)^2}$$

Formula (25) shows that the external estimate $Est[E(D_2(E_\lambda(\mathbf{q} \mid \mathbf{e}_n),\mathbf{q})]$ of the prospect of failure of E_λ in the 'actual world' is *identical* to the prospect of failure $E(D_2(E_\lambda(\mathbf{q} \mid \mathbf{e}_n),q) \mid q)$ of E_λ in a 'possible world' q where $G(q) = Est[G(\mathbf{q})]$ and $D_2(\gamma^\circ,q) = Est(D_2(\gamma^\circ,\mathbf{q})]$ (see (23)). From this it follows that the (unique) V_2-optimum λ-value minimizing $Est[E(D_2(E_\lambda(\mathbf{q} \mid \mathbf{e}_n),\mathbf{q})]$, which may be referred to as λ_2°, is *identical* to the λ-value λ_1° minimizing $E(D_2(E_\lambda(\mathbf{q} \mid \mathbf{e}_n),q) \mid q)$ (see (24)): $\lambda_2^\circ = \lambda_1^\circ = Est[G(\mathbf{q})]/Est[D_2(\gamma^\circ,\mathbf{q})]$.

This signifies that the sophisticated V-solution V_2 and the naive V-solution V_1 are two different formulations of a unique V-solution to EPO for GD-contexts. The *V-optimum λ-value* $\lambda_1^\circ = \lambda_2^\circ$ may be referred to as λ_V°.

8.6 THE EQUIVALENCE BETWEEN THE V-SOLUTION AND THE CC-SOLUTION TO EPO

Although, as remarked above, the equality $\lambda_{CC} = \lambda_V^\circ$ may appear somewhat fortuitous, it can be shown that it is a necessary consequence of the fact that the V- and the CC-solutions to EPO are, contrary to appearances, *logically equivalent*. An outline of my argument is as follows:

(i) the V_2-optimum λ-value λ_2° (= λ_V°) is defined as the (unique) λ-value minimizing $Est[E(D_2(E_\lambda(\mathbf{q} \mid \mathbf{e}_n),\mathbf{q})]$ within the set of Dir(γ°,λ)-based M-estimators E_λ (see above);

(ii) the *definition* of the CC-solution to EPO *implies* that $Est[E(D_2(E_\lambda(\mathbf{q} \mid \mathbf{e}_n),\mathbf{q})]$ is also minimized by the CC-optimum λ-value λ_{CC};

(iii) therefore λ_{CC} is necessarily identical to λ_V°.

The core of the argument is represented by point (ii) which can be proved using theorem (4.12) which concerns the immodesty of any prior distribution on $\mathbf{q}$.

Before developing the argument in detail, three lemmas must be proved.

Given a value q of $\mathbf{q}$, an experiment $\mathbf{e}_n$, a prior Dir(γ°,λ') on $\mathbf{q}$, the expected (w.r.t. $P_q(e_n)$) quadratic distance between the Dir(γ°,λ')-based M-estimate $E_{\lambda'}(\mathbf{q} \mid \mathbf{e}_n)$ and q is referred to as $E(D_2(E_{\lambda'}(\mathbf{q} \mid \mathbf{e}_n),q) \mid q)$ (see Section 4). Moreover, given a prior Dir(γ°,λ) on $\mathbf{q}$, the expected (w.r.t. Dir(γ°,λ)) value of $E(D_2(E_{\lambda'}(\mathbf{q} \mid \mathbf{e}_n),\mathbf{q}) \mid \mathbf{q})$ will be referred to as

$E_\lambda[E(D_2(E_{\lambda'}(\mathbf{q} \mid \mathbf{e_n}),\mathbf{q}) \mid \mathbf{q})]$. Lastly, the expected value of the quadratic distance $D_2(E_{\lambda'}(\mathbf{q} \mid \mathbf{e_n}),\mathbf{q})$ - w.r.t. the joint probability distribution on $(\mathbf{q},\mathbf{e_n})$ derived from Dir($\gamma°,\lambda$) and the likelihood function $p(e_n/q) = P_q(e_n)$) - is referred to as $E_\lambda[D_2(E_{\lambda'}(\mathbf{q} \mid \mathbf{e_n}),\mathbf{q})]$.

The first lemma states the following equality:

(26) $$E_\lambda[D_2(E_{\lambda'}(\mathbf{q} \mid \mathbf{e_n}),\mathbf{q})] = E_\lambda[E(D_2(E_{\lambda'}(\mathbf{q} \mid \mathbf{e_n}),\mathbf{q}) \mid \mathbf{q})]$$

Equality (26) is a particular case of (3.19).[15]

Given the CC-optimum prior Dir($\gamma°,\lambda_{CC}$) and the corresponding *M*-estimator E_{CC}, the second lemma states that:

(27) $$E_{CC}[D_2(E_{CC}(\mathbf{q} \mid \mathbf{e_n}),\mathbf{q})] \leq E_{CC}[D_2(E_{\lambda'}(\mathbf{q} \mid \mathbf{e_n}),\mathbf{q})]$$

where $E_{CC}[D_2(E_{CC}(\mathbf{q} \mid \mathbf{e_n}),\mathbf{q})]$ and $E_{CC}[D_2(E_{\lambda'}(\mathbf{q} \mid \mathbf{e_n}),\mathbf{q})]$ are the expected values of $D_2(E_{CC}(\mathbf{q} \mid \mathbf{e_n}),\mathbf{q})$ and $D_2(E_{\lambda'}(\mathbf{q} \mid \mathbf{e_n}),\mathbf{q})$ calculated w.r.t. the joint probability distribution on $(\mathbf{q},\mathbf{e_n})$ derived from Dir($\gamma°,\lambda_{CC}$) and $p(e_n/q) = P_q(e_n)$.

Inequality (27) is a particular case of (4.12). The intuitive meaning of (27) is that Dir($\gamma°,\lambda_{CC}$) is *immodest* as is any other prior distribution on **q** (cf. (4.10)).

Lastly, the third lemma can be stated thus:

(28) $$E_\lambda[E(D_2(E_{\lambda'}(\mathbf{q} \mid \mathbf{e_n}),\mathbf{q}) \mid \mathbf{q})] = \frac{nE_\lambda[G(\mathbf{q})] + (\lambda')^2 E_\lambda[D_2(\gamma°,\mathbf{q})]}{(n + \lambda')^2}$$

where $E_\lambda[G(\mathbf{q})]$ and $E_\lambda[D_2(\gamma°,\mathbf{q})]$ denote the expected (w.r.t. Dir($\gamma°,\lambda$)) values of $G(\mathbf{q})$ and $D_2(\gamma°,\mathbf{q})$.

Proof of (28): By replacing λ with λ' in formula (20) the equality $ED_2(E_{\lambda'}(\mathbf{q} \mid \mathbf{e_n}),q) = [nG(q) + (\lambda')^2 D_2(\gamma°,q)]/(n + \lambda')^2$ is obtained. From this equality and from the property (3.16.a) of expectations, it follows that $E_\lambda[E(D_2(E_{\lambda'}(\mathbf{q} \mid \mathbf{e_n}),\mathbf{q}) \mid \mathbf{q})] = [nE_\lambda[G(\mathbf{q})] + (\lambda')^2 E_\lambda(D_2(\gamma°,\mathbf{q})]/(n + \lambda')^2$, *Q.E.D.*

My argument develops along these lines.

It follows from (28) that the expected (w.r.t. Dir($\gamma°,\lambda_{CC}$)) value $E_{CC}[E(D_2(E_{\lambda'}(\mathbf{q} \mid \mathbf{e_n}),\mathbf{q}) \mid \mathbf{q})]$ of the prospect of failure $E(D_2(E_{\lambda'}(\mathbf{q} \mid \mathbf{e_n}),\mathbf{q}) \mid \mathbf{q})$ of an arbitrary Dir($\gamma°,\lambda'$)-based *M*-estimator $E_{\lambda'}$ is given by:

$$(29) \qquad E_{CC}[E(D_2(E_{\lambda'}(\mathbf{q} \mid \mathbf{e}_n),\mathbf{q}) \mid \mathbf{q})] = \frac{nE_{CC}[G(\mathbf{q})] + (\lambda')^2 E_{CC}[D_2(\gamma^\circ,\mathbf{q})]}{(n + \lambda')^2}$$

where E_{CC} denotes the expectation w.r.t. Dir($\gamma^\circ,\lambda_{CC}$).

Remember that the definition of the CC-solution implies that $E_{CC}[G(\mathbf{q})] = Est[G(\mathbf{q})]$ and $E_{CC}[D_2(\gamma^\circ,\mathbf{q})] = Est[D_2(\gamma^\circ,\mathbf{q})]$ (see Section 3). Hence, equality (29) can be rewritten as follows:

$$(30) \qquad E_{CC}[E(D_2(E_{\lambda'}(\mathbf{q} \mid \mathbf{e}_n),\mathbf{q}) \mid \mathbf{q})] = \frac{nEst(G(\mathbf{q}) + (\lambda')^2 Est[D_2(\gamma^\circ,\mathbf{q})]}{(n + \lambda')^2}$$

By replacing λ with λ' in formula (25) the external estimate $Est[E(D_2(E_{\lambda'}(\mathbf{q} \mid \mathbf{e}_n),\mathbf{q})]$ of the prospect of failure $E(D_2(E_{\lambda'}(\mathbf{q} \mid \mathbf{e}_n),\mathbf{q})$ of $E_{\lambda'}$ in the 'actual world' is obtained thus:

$$(31) \qquad Est[E(D_2(E_{\lambda'}(\mathbf{q} \mid \mathbf{e}_n),\mathbf{q})] = \frac{nEst[G(\mathbf{q})] + (\lambda')^2 Est[D_2(\gamma^\circ,\mathbf{q})]}{(n + \lambda')^2}$$

Formulae (30) and (31) imply that:

$$(32) \qquad E_{CC}[E(D_2(E_{\lambda'}(\mathbf{q} \mid \mathbf{e}_n),\mathbf{q}) \mid \mathbf{q})] = Est[E(D_2(E_{\lambda'}(\mathbf{q} \mid \mathbf{e}_n),\mathbf{q})]$$

Note that two particular cases of (26) are given by:

$$E_{CC}[D_2(E_{CC}(\mathbf{q} \mid \mathbf{e}_n),\mathbf{q})] = E_{CC}[E(D_2(E_{CC}(\mathbf{q} \mid \mathbf{e}_n),\mathbf{q}) \mid \mathbf{q})]$$

and

$$E_{CC}[D_2(E_{\lambda'}(\mathbf{q} \mid \mathbf{e}_n),\mathbf{q})] = E_{CC}[E(D_2(E_{\lambda'}(\mathbf{q} \mid \mathbf{e}_n),\mathbf{q}) \mid \mathbf{q})].$$

From these equalities and (27) it follows that:

$$(33) \qquad E_{CC}[E(D_2(E_{CC}(\mathbf{q} \mid \mathbf{e}_n),\mathbf{q}) \mid \mathbf{q})] \leq E_{CC}[E(D_2(E_{\lambda'}(\mathbf{q} \mid \mathbf{e}_n),\mathbf{q}) \mid \mathbf{q})]$$

Inequality (33) signifies that $E_{CC}[E(D_2(E_{\lambda'}(\mathbf{q} \mid \mathbf{e}_n),\mathbf{q}) \mid \mathbf{q})]$ is minimized by $\lambda' = \lambda_{CC}$. Given (32), $Est[E(D_2(E_{\lambda'}(\mathbf{q} \mid \mathbf{e}_n),\mathbf{q})]$ is also minimized by $\lambda' = \lambda_{CC}$. Recalling that the V_2-optimum λ-value λ_2° (= λ_V°) is defined as the (unique) λ-value minimizing $Est[E(D_2(E_{\lambda'}(\mathbf{q} \mid \mathbf{e}_n),\mathbf{q})]$ (see the remarks below (25)), this implies that λ_{CC} is necessarily identical to λ_V°.

8.7 CARNAP'S OPTIMUM INDUCTIVE METHODS

Given the equivalence - mentioned in the introductory paragraphs - between a Dirichlet prior Dir(γ,λ) and the corresponding GC-system (γ,λ), my V-solution to LPO for DIR(γ) (see Section 4) can be reinterpreted as a V-solution to LPO for the set GC(γ) of GC-systems with prior vector γ.

Two principal objectives of this section are: (i) to provide a 'verisimilitude re-interpretation' of Carnap's approach (1952) to optimum C-systems and (ii) to show that my V-solution to LPO for GC-system can be seen as a generalization of Carnap's solution to LPO for C-systems.

My V-solution equates the optimum λ-value to $\lambda^{\Delta} = G(q)/D_2(\gamma,q)$ (see (21)). In particular, regarding the set DIR(**1/k**) of *symmetrical Dirichlet distributions* - or, equivalently, the set GC(**1/k**) of *C-systems* - the value of λ^{Δ} is derived from (21) by replacing γ with the indifference vector **1/k**=(1/*k*,...,1/*k*): $\lambda^{\Delta} = G(q)/D_2(\mathbf{1/k},q)$. Given the equalities $G(q) = 1 - \Sigma q_i^2 = 1 - C(q)$ and $D_2(\mathbf{1/k},q) = \Sigma(q_i - 1/k)^2 = \Sigma q_i^2 - 1/k = C(q) - 1/k$, the formula for λ^{Δ} can be rewritten as follows:

$$(34) \qquad \lambda^{\Delta} = \frac{(1 - \Sigma\, q_i^2)}{(\Sigma\, q_i^2 - 1/k)} = \frac{1 - C(q)}{C(q) - 1/k}$$

Formula (34) is identical to the formula for the optimum λ-value found by Carnap (1952, p. 69).[16] Below it is shown that this identity derives from the conceptual relationships existing between my approach to LPO and that of Carnap.[17]

Let $E_\lambda(\mathbf{q_i}/e_n) = p_\lambda(Q_i/e_n)$ be the expected value of the objective probability $\mathbf{q_i}$ of Q_i in a given population, where $p_\lambda(Q_i/e_n)$ is a special value of the C-system (**1/k**,λ) and e_n is a possible result of the experiment $\mathbf{e_n}$. Carnap refers to the estimates $E_\lambda(\mathbf{q_i}/e_n)$, relative to the possible results of $\mathbf{e_n}$, and proposes the following four step strategy to identify the (logically) optimum C-system:

(S.1) A measure $ser(E_\lambda(\mathbf{q_i} \mid e_n),q_i)$ of the *seriousness of the error* made by $E_\lambda(\mathbf{q_i} \mid e_n)$ w.r.t. a value q of $\mathbf{q}$ is established.

(S.2) The (*local*) Q_i*-prospect of failure* of (**1/k**,λ) in q, relative to $\mathbf{e_n}$, is defined as the expected (w.r.t. $P_q(e_n)$) value $E(ser(E_\lambda(\mathbf{q_i} \mid \mathbf{e_n}),q_i) \mid q)$ of $ser(E_\lambda(\mathbf{q_i} \mid \mathbf{e_n}),q_i)$.

(S.3) The (*global*) *prospect of failure* of (**1/k**,λ) in q, relative to $\mathbf{e_n}$, is defined as the arithmetic average $\Sigma E(ser(E_\lambda(\mathbf{q_i} \mid \mathbf{e_n}),q_i) \mid q)/k$ of the k Q_i-prospects of failure.

(S.4) The *optimum λ-value* in q, relative to $\mathbf{e}_n$, is defined as the λ-value minimizing the global prospect of failure.

The following 'verisimilitude re-interpretation' of Carnap's strategy is here suggested. The seriousness of the error $ser(E_\lambda(\mathbf{q}_i \mid e_n), q_i)$ can be seen as a measure of the *local distance* between the estimate $E_\lambda(\mathbf{q}_i \mid e_n)$ and the true value q_i of $\mathbf{q}_i$. Accordingly, the optimum λ-value may be seen as the λ-value minimizing the arithmetic average $\Sigma E(ser(E_\lambda(\mathbf{q}_i \mid \mathbf{e}_n), q_i) \mid q)/k$ of the expected local distances $E(ser(E_\lambda(\mathbf{q}_i \mid \mathbf{e}_n), q_i) \mid q)$.

To apply Carnap's strategy, (S.1) should be specified by selecting an appropriate definition of $ser(E_\lambda(\mathbf{q}_i \mid e_n), q_i)$. In particular, Carnap (*ibid.*, pp. 61-62) defines $ser(E_\lambda(\mathbf{q}_i \mid e_n), q_i)$ as the *square error* of $E_\lambda(\mathbf{q}_i \mid e_n)$ relative to the value q_i of $\mathbf{q}_i$:[18]

(35) $$ser(E_\lambda(\mathbf{q}_i \mid e_n), q_i) = (E_\lambda(\mathbf{q}_i \mid e_n) - q_i)^2$$

It follows from (35) - in combination with (S.2), (S.3) and (3.16.b) - that the prospect of failure $\Sigma E(ser(E_\lambda(\mathbf{q}_i \mid \mathbf{e}_n), q_i) \mid q)/k$ in (S.3) is equal to the *expected average square error*:

(36) $$\Sigma\, E(ser(E_\lambda(\mathbf{q}_i \mid \mathbf{e}_n), q_i) \mid q)/k = E(\Sigma\, (E_\lambda(\mathbf{q}_i \mid \mathbf{e}_n) - q_i)^2/k \mid q)$$

Note that the average square error $\Sigma(E_\lambda(\mathbf{q}_i \mid e_n) - q_i)^2/k$ is a simple linear transformation of the quadratic distance $D_2(E_\lambda(\mathbf{q} \mid e_n), q) = \Sigma(E_\lambda(\mathbf{q}_i \mid e_n) - q_i)^2$ used in my V-approach to LPO. Hence, $E(\Sigma(E_\lambda(\mathbf{q}_i \mid \mathbf{e}_n) - q_i)^2/k \mid q)$ is a simple linear transformation of $E(D_2(E_\lambda(\mathbf{q} \mid \mathbf{e}_n), q \mid q)$: accordingly both quantities are minimized by $\lambda^\Delta = (1 - \Sigma q_i^2)/(\Sigma q_i^2 - 1/k)$. This signifies that my V-solution to LPO for GC-systems is logically equivalent to a *generalization* of Carnap's solution to LPO for C-systems.[19]

However, there are also some important differences between my V-approach to LPO and that of Carnap.

Firstly, while my V-solution to LPO for the set of symmetrical *Dirichlet distributions* and the corresponding set of C-systems is worked out by making a substantial usage of the mathematical properties of Dirichlet distributions, Carnap fails to mention Dirichlet distributions at all.

Secondly, while my V-solution to LPO is based on the *explicit* adoption of a verisimilitude view of the aims of science, Carnap does not state any formal notion of verisimilitude and, as a matter of fact, this term does not appear in *The Continuum of Inductive Methods* (Carnap, 1952).[20]

CHAPTER 9

THE CONTEXTUAL APPROACH TO EPO: COMPARISONS WITH OTHER VIEWS

In this chapter some features of the *contextual view* of optimum inductive methods (optimum prior distributions) will be elucidated by comparing this view with others.

In Section 1 it will be seen that, underlying Carnap's oscillations between subjectivism and apriorism, there is a constant assumption, i.e., his *universalistic view* according to which a person at any given time should select a *unique* epistemically optimum inductive method for *all* those scientific and practical problems which he has to deal with.

In Section 2 the solutions to EPO for C-systems proposed by Lewis (1972) and Kuipers (1986) will be examined. They are seen as different formulations of a *hyperempiricist view* where the experimental evidence e_n *alone* is sufficient to identify the optimum C-system.

In Section 3 the *presupposition view* of induction will be considered. According to this view the selection of the inductive method should reflect certain presuppositions regarding the degree of order of the 'sector of the world' under examination.

Finally, in Section 4 the *verisimilitude view*, according to which an inductive method should help to approach the truth, will be examined.

From the above mentioned comparisons it results that the contextual view differs greatly from both the universalistic view and the hyperempiricist view, while it can be seen as a development and application of both the presupposition and verisimilitude views.

9.1 THE UNIVERSALISTIC VIEW

Regarding the choice of the inductive method Carnap (1952, p. 53), points out that

> ... the question as to which of the available methods a man *X* ought to choose ... is fundamentally not a theoretical question. A possible answer to a theoretical question is an

assertion; as such it can be judged as true or false, and, if it is true, it demands the assent of all. Here, however, the answer consists in a *practical decision* to be made by *X*. A decision cannot be judged as true or false but only as *more or less adequate, that is, suitable for given purposes*. However, the adequacy of the choice depends, of course, on many theoretical results concerning the properties of the various inductive methods; and therefore the theoretical results may influence the decision. Nevertheless, *the decision itself still remains a practical matter*, a matter of *X* making up his mind, *like choosing an instrument for a certain kind of work.* [Author's italics]

Carnap (*ibid.*, pp. 54-55) develops his comparison between inductive methods and instruments as follows:

Suppose that *X* has chosen a certain inductive method and used it during a certain period for the inductive problems which occurred. If, in view of the services this method has given him, he is not satisfied with it, he may at any time abandon it and adopt another method which seems to him preferable. *This is not the same as a change in method from problem to problem. Once he adopts the new method, he will apply it to all inductive problems, problems of confirmation for all kinds of hypotheses ... One inductive method is here envisaged as covering all inductive problems. ... An inductive method is ... an instrument for the task of constructing a picture of the world on the basis of observational data* and especially of forming expectations of future events as a guidance for practical conduct. *X may change this instrument just as he changes a saw or an automobile, and for similar reasons.* If *X*, after using his car for some time, is no longer satisfied with it, he will consider taking another one, provided that he finds one that seems to him preferable. Relevant points of view for his preference might be: performance, economy, aesthetic satisfaction, and others. Similarly, after working with a particular inductive method for a time, he may not be quite satisfied and therefore look around for another method. He will take into consideration *the performance of a method, that is, the values it supplies and their relation to later empirical results, e.g. the truth-frequency of predictions and the error of estimates*; further, the economy in use, measured by the simplicity of the calculations required; maybe also aesthetic features, like the logical elegance of the definitions and rules involved. ... Here, as anywhere else, life is a process of never ending adjustment; there are no absolutes, neither absolutely certain knowledge about the world nor absolutely perfect methods of working in the world. [Author's italics]

Some remarks are called for. Firstly, the choice of an inductive method is considered by Carnap as a "*practical decision*". In one sense, Carnap is right since the adoption of an inductive method may have many practical consequences: for instance, the predictive probabilities derived from an inductive method can be used, in combination with suitable utilities, for making several kinds of practical decisions. In other words, the selection of an inductive method can be seen as a kind of far-reaching practical 'meta-decision', since it leads to the adoption of a general strategy for making different types of decisions.

However, I would argue that the ultimate roots of such a meta-decision are cognitive. In fact, as illustrated by the verisimilitude approach to EPO, the choice of an inductive method can be seen as a *cognitive decision* aimed at achieving a particular *cognitive utility*, i.e., approximation to the truth.

On the other hand, in the above quotations, Carnap himself appears to recognize that at least some of the "given purposes" w.r.t. which an inductive method is judged as "more or less adequate" are *cognitive purposes* and that, consequently, an inductive method is not only a practical instrument but also, as it were, a cognitive instrument. Carnap, indeed, points out that "an inductive method is ... *an instrument for the task of constructing a picture of the world* on the basis of observational data" and that the "performance" of an inductive method depends on "the values it supplies and their relation to later empirical results, e.g. the truth-frequency of predictions and the error of estimates". It seems quite clear that Carnap is specifically referring here to the *cognitive* performance of a given inductive method and that, in addition, this performance is somehow related to its effectiveness in approaching the truth.

Moreover, Carnap seems to suggest that the 'past performance' of a given inductive method is the principal basis for deciding between retaining such a method or adopting another one. This view could be called 'empiricist' since the adoption of an inductive method depends essentially on the empirical evaluation of its *past* performance. The *empiricist approach* to EPO represents a possible alternative to the subjectivistic and aprioristic approaches advocated by Carnap in other works.

However, there is a constant assumption underlying Carnap's oscillations among apriorism, subjectivism and empiricism. This assumption, which might be called *universalistic*, presumes that any subject *X*, at any time *t*, can select a *unique* epistemically optimum inductive method which is universally applicable to *all* scientific inquiries and practical problems facing *X* at *t*.

Carnap's universalistic approach is precisely the opposite of our contextual approach to EPO where the selection of the optimum inductive method depends on the specific cognitive context.

EPO corresponds approximately to Carnap's "problem of the success [of the different C-systems] in the actual universe" (1952, p. 75). Carnap is aware that this problem is different from LPO and that its solution cannot be directly obtained from a solution to LPO. In fact, "since an observer knows at any time only a sample, not the whole universe, he cannot determine the optimum inductive method for the actual universe" (Carnap, 1952, p. 2).

On the other hand, it might be suggested that *X* can estimate, in some way, the success of the different C-systems in the actual universe and that, using such estimates, he can also 'estimate' the optimum value λ^{A}. Contrary to this suggestion, Carnap (*ibid.*, pp. 59-60) maintains that:

Questions concerning the success of a given inductive method in the actual world would be of a factual, nonlogical nature. And if they concerned not merely that part of the world which is known to us by past observation but also a part or the whole of the future, then the answer could be given with certainty only after all observations reports were in, if that were ever possible. *And if our question concerned not the actual success but the probability of success or an estimate of success, then it would make sense only on the basis of a chosen inductive method.* [Author's italics]

The last point would signify that the success (failure) of any inductive method in the actual universe cannot be estimated without *presupposing* a given inductive method λ.[1] In fact Carnap (*ibid.*, p.60) remarks that:

... any judgment about the success of an inductive method in the total actual world ... is obviously impossible from an inductively neutral standpoint.

I believe there are good grounds for opposing Carnap's statement. In fact, from the external estimates $Est(\mathbf{q})$, $Est[G(\mathbf{q})]$, and $Est[D_2(\gamma^{\circ},\mathbf{q})]$ made in GD-contexts (see Chapter 8.2) an external estimate of the cognitive failure of any GC-system in the actual universe can be derived (see (8.25)). It seems clear that, while this estimate *is* an inductive estimate (based on the background knowledge *BK*), it *is not* a λ-based inductive estimate. To use Carnap's words, it is a "judgment about the success of [a GC-system] in the total actual world" which is made "from an inductively neutral standpoint".

Given the *inductive neutrality* of the above external estimates they would appear to provide an 'unprejudiced criterium' for selecting the epistemically optimum GC-system.

Carnap (1950, pp. 177-182) also rejects the idea that selection of an inductive method might be justified using the so-called *principle of uniformity*. For instance, Carnap (*ibid.*, p. 180) considers the following probabilistic formulation for the principle of uniformity:

On the basis of the available evidence, the *estimate* of the degree of uniformity of the world is high.

Carnap assumes that the degree of uniformity of the world cannot be estimated without *presupposing* a given inductive method; hence the principle of uniformity cannot be used to justify the choice of the inductive method.

However I believe that Carnap's assumption is mistaken since - at least in an important type of multinomial context, i.e. GD-contexts - inductively neutral estimates $Est[G(\mathbf{q})]$ of the degree of disorder in the "actual world" can be made (see Chapter 8.2). Moreover, contrary to Carnap's conclusion, I hold that such estimates can be used for the choice of the optimum GC-system (see Chapter 8, Sections 3-5).

9.2 THE HYPEREMPIRICIST VIEW

Lewis's approach to EPO

According to Lewis's solution to EPO for C-systems (1971) the experimental evidence e_n *alone* is sufficient to identify the optimum C-system.

In Lewis's paper the *requirement of immodesty* plays a fundamental role (see Chapter 4.2). According to Lewis (1971, p. 56):

The requirement of immodesty will not help you much in choosing an inductive method unless few of the otherwise adequate methods are immodest. I might expect all methods to be immodest; in that case, it will get you nowhere to require immodesty as a condition of adequacy. How many methods are immodest?

To answer this last question the class of competing inductive methods one wishes to choose from must be specified. For instance, given the class of C-systems, a C-system λ is *immodest*, on the evidence e_n, "iff its estimate, on $[e_n]$, of its own accuracy is higher than its estimate, on $[e_n]$, of the accuracy of any rival method [λ']" (Lewis, 1974, p. 84).

As has already been pointed out (see Chapter 4.2), Lewis (1971) fails to understand that all C-systems and, more generally, all exchangeable inductive systems are immodest. On the contrary - due to a subtle conceptual mistake in formulating his equations - Lewis (*ibid.*, p. 59) believes to have proved the surprising theorem that, for any evidence e_n, exactly *one* C-system is immodest.[2]

Lewis (*ibid.*, pp. 59-60) points out that his 'theorem' provides a complete solution to EPO:

Our new condition of adequacy, immodesty, and the conditions of adequacy that restrict you to the [C-systems] are enough to solve completely your problem of choosing an inductive method. You need only choose the one remaining adequate method. The conditions of adequacy thereby determine the degrees to which you should believe all propositions of interest to you, given total evidence consisting of a complete description of a sample.

Although the 'immodesty-based' solution to EPO proposed by Lewis (1971) is erroneous, it may be asked whether Lewis's more general 'hyperempiricist program' - according to which a suitable set of conditions of adequacy could "determine the degrees to which you should believe all propositions of interest to you, given total evidence consisting of a complete description of a sample" (Lewis, 1971, p. 60) - can be carried out.

I suspect that the hyperempiricist program is undermined by some intrinsic difficulties as will be shown after examining Kuipers's hyperempiricist proposal (1986). Here I only want to suggest that, contrary to initial impressions, the hyperempiricist approach is a (disguised) form of apriorism. Indeed, to all appearances, Lewis's approach to EPO is "the other extreme" w.r.t. Carnap's apriorism, as pointed out, for instance, by Pietarinen (1974, p. 198):

> Carnap's latest view is that there is (ideally) only one rational λ-method, the choice of which is not based on observed facts but only on *a priori* considerations ... let this be $[\lambda_R]$. Then the actual belief in a hypothesis h at any two points of time T_1 and T_2 is $[p(h/e_{T1})]$ and $[p(h/e_{T2})]$ respectively, where e_{Ti} is the total evidence at T_i. ... Lewis takes the other extreme: he makes the choice of a λ-method depend essentially on the total evidence e_{Ti}. At T_1 and T_2 the rational belief in h is $[p_{\lambda 1}(h/e_{T1})]$ and $[p_{\lambda 2}(h/e_{T2})]$ respectively, where the values λ_1 and λ_2 depend on e_{T1} and e_{T2}.

However, careful scrutiny reveals that Lewis's "other extreme" is simply a modified form of apriorism. Indeed, according to Carnap's 'traditional' aprioristic approach to EPO the prior distribution p should be chosen on the grounds of appropriate a priori conditions of adequacy. Hence, the posterior probabilities $p(Q_i/e_n)$ - derived from p and e_n by Bayes's theorem - should depend only on these a priori conditions and e_n. But, after all, the same happens within the hyperempiricist approach where the posterior probabilities $p(Q_i/e_n)$ are determined only by the experimental evidence e_n and certain a priori conditions of adequacy, such as the principle of immodesty.

The only difference between the two approaches is that in the 'traditional' aprioristic approach the prior distribution p is determined *before* observing e_n and only after is it updated relative to e_n, while in the 'modified' aprioristic approach suggested by Lewis and Kuipers p is determined *after* obtaining e_n - and, so to say, from e_n - by suitable a priori conditions of adequacy.[3]

Both Carnap's and the Lewis-Kuipers approaches attempt to determine the posterior probabilities $p(Q_i/e_n)$ using *only* experimental evidence e_n. On the contrary, in the contextual approach to EPO as proposed herein the background information concerning a given population (or process) available to researchers *before* commencing collection of experimental data plays a crucial role in determining the prior probabilities and, consequently, the posterior probabilities $p(Q_i/e_n)$.[4]

Kuipers's approach to EPO

An interesting formulation of the hyperempiricist program is given by Kuipers (1986). Kuipers (1986, p. 39) assumes that, although the logically optimum value λ^{Δ} for investigating a given process *Ex* cannot be known since it depends on the (typically unknown) degree of homogeneity $C(\mathbf{q})$ of *Ex*, it is possible to estimate λ^{Δ} on the basis of e_n and, therefore, this estimated value of λ^{Δ} can be adopted as the epistemically optimum λ-value.

Kuipers proposes two different procedures to 'estimate' λ^{Δ}: an 'estimation by a limit-process' and a 'one-step estimation'.

Using the first procedure the e_n-based estimate of λ^{Δ} is made thus (*ibid.*, p. 39):

(i) starting from an arbitrary (finite) value λ_0 of λ, the value of the unknown $C(\mathbf{q})$ is estimated,
(ii) the estimated value of $C(\mathbf{q})$ is used to obtain a 'tentative estimate' λ_1 of λ^{Δ},
(iii) the estimated value λ_1 of λ^{Δ} is used to restart this cycle and so on for an infinite number of times.

Kuipers succeeds in proving that the above cycle is a limit-process. In other words, for any experimental evidence e_n and for any 'starting' λ-value λ_0, the sequence of 'tentative estimates' $\lambda_1,...,\lambda_m,...$ of the cycle has a limit - let us call it "λ_{Ku}" - which can be seen as the 'ultimate estimate' of λ^{Δ} (*ibid.*, pp. 40-41).

Kuipers (*ibid.*, pp. 39-40) characterizes the steps (i)-(ii) of the above procedure as follows. Given the starting value λ_0, the 'corresponding' estimate $est_{\lambda 0}[C(\mathbf{q})]$ of $C(\mathbf{q})$ is drawn from the inductive probabilities $p_{\lambda 0}(Q_i/e_n)$ by the following formula:

$$(1)\quad est_{\lambda 0}[C(\mathbf{q})] = \Sigma\, [p_{\lambda 0}(Q_i/e_n)]^2$$

Afterwards λ_1 is defined as the λ-value that would be the logically optimum value if the estimate $est_{\lambda 0}[C(\mathbf{q})]$ were correct:

$$(2)\quad \lambda_1 = \frac{1 - est_{\lambda 0}[C(\mathbf{q})]}{est_{\lambda 0}[C(\mathbf{q})] - 1/k}$$

At any moment in the cycle the same procedure should be applied when an estimate λ_{m-1} is used to establish the corresponding estimate $est_{\lambda_{m-1}}[C(\mathbf{q})]$ and the next estimate λ_m.[5]

A neat feature of the above limit-process is the restricted role played by the starting λ-value λ_0 in establishing the epistemically optimum value λ_{Ku}. In fact, Kuipers proves that, for any evidence e_n, there are no more than two possible non-extreme results of the limit-process.[6]

This means that the effects of the inevitably arbitrary choice of the starting λ-value λ_0 are to a large extent eliminated by the limit-process and consequently the freedom of researchers in selecting the epistemically optimum λ-value is restricted, at most, to the choice between the two values of λ. Hence, Kuipers's first proposal can be considered an 'almost complete' solution to EPO for C-systems.

Let us now consider one interesting feature of the above limit-process overlooked by Kuipers himself. Taking one of the possible limit-values λ_{Ku} as the starting λ-value λ_0 of the cycle, then the cycle reaches its final result immediately since in this case $\lambda_{Ku} = \lambda_0 = \lambda_1 = \lambda_2 = ... = \lambda_m = ...$.

Therefore I would propose the above solution to EPO be restated using a concept of immodesty analogous to, but different from, that introduced by Lewis. Kuipers's limit-process can be seen, indeed, as a tool for arriving at an 'immodest' inductive method λ_{Ku} through an infinite sequence of 'modest' inductive methods $\lambda_0,\lambda_1,...,\lambda_m,...$, starting from an arbitrary method λ_0. Each method $\lambda_{m\text{-}1}$ in the sequence is modest in the sense that - through the inductive probabilities $p_{\lambda_{m\text{-}1}}(Q_i\,/e_n)$ and the 'corresponding' estimates $est_{\lambda_{m\text{-}1}}[C(\mathbf{q})]$ of $C(\mathbf{q})$ - it leads to a different and more appropriate estimate λ_m of the optimum value λ^{Δ}. However, 'at the end' of this infinite cycle an 'immodest' method λ_{Ku} is obtained indicating itself as the most appropriate estimate of λ^{Δ}.[7]

Hence, Kuipers's first solution to EPO for C-systems can be reformulated as follows: given the evidence e_n, the immodest method λ_{Ku} which indicates itself as the most appropriate estimate of λ^{Δ} should be selected as the epistemically optimum inductive method.

Note that, while *any* inductive method is immodest relative to my notion of immodesty (see 4.12), only one method for each e_n is immodest if Kuipers's notion of immodesty is applied. The highly selective character of Kuipers's notion of immodesty depends on the specific procedure he proposes for passing from $\lambda_{m\text{-}1}$ to λ_m.

Let us accept, for the sake of argument, Kuipers's two-step procedure (i)-(ii) for determining the first tentative estimate λ_1 of λ^{Δ} starting from a given evidence e_n and an initial λ-value λ_0. Yet it could be argued that formula (1) does not represent the best procedure for determining the λ_0-based estimate $est_{\lambda_0}[C(\mathbf{q})]$ requested in step (i). Initially (1) seems plausible enough since $p_0(Q_i\,/e_n) = E_{\lambda_0}(\mathbf{q}_i)$ appears an appropriate λ_0-based estimate of $\mathbf{q}_i$ and, consequently, it might be thought that an appropriate estimate $est_{\lambda 0}[C(\mathbf{q})]$ of $C(\mathbf{q})$ may be obtained by replacing all the $\mathbf{q}_i$ in $C(\mathbf{q}) = \Sigma \mathbf{q}_i^2$ with the corresponding estimates $E_{\lambda 0}(\mathbf{q}_i) = p_0(Q_i/e_n)$.

However, from a Bayesian point of view, (1) would not appear acceptable. In fact, while $p_0(Q_i/e_n)$ *is* an appropriate λ_0-based Bayesian estimate of $\mathbf{q}_i$ since it is equal to the expected value $E_{\lambda 0}(\mathbf{q}_i)$ of $\mathbf{q}_i$, the estimate $est_{\lambda 0}[C(\mathbf{q})] = \Sigma[p_{\lambda 0}(Q_i/e_n)]^2$ proposed in (1) *is not* an appropriate λ_0-based Bayesian estimate of $C(\mathbf{q})$. In fact $\Sigma[p(Q_i/e_n)]^2$ is different from (and typically higher than) the expected value $est_{\lambda 0}(C(\mathbf{q}) \mid e_n) = \Sigma(p_{\lambda 0}(Q_i Q_i/e_n)$ which is the usual λ_0-based Bayesian estimate of $C(\mathbf{q})$.[8]

This argument assumes that a C-system λ, being equivalent to a prior symmetrical Dirichlet on $\mathbf{q}$, should be treated as any other 'Bayesian method' and, accordingly, a λ-based estimate of a given quantity should be made by applying the *M*-estimator or another *Bayesian* procedure. Against this assumption, however, it might be argued that using non-Bayesian λ-based estimates for selecting the optimum C-system is a perfectly legitimate 'Bayes/non-Bayes compromise' of the type welcomed by Good (1965, pp. 28 and 34-41; 1983a, pp. 100-103).

A 'Bayes/non-Bayes interpretation' applies clearly to Kuipers's second solution to EPO for C-systems (*ibid.*, pp. 43-44). Indeed, the 'one-step estimation' of λ^{Δ} proposed does not require any starting method λ_0 to estimate the value of $C(\mathbf{q})$ relative to a given evidence e_n. For this purpose, one of the usual orthodox (non-Bayesian) statistical techniques is suggested such as the following simple estimation procedure: $est(C(\mathbf{q}) \mid e_n) = \Sigma(n_i/n)^2$. After establishing $est(C(\mathbf{q}) \mid e_n)$, the epistemically optimum inductive method is equated to the method λ^* which would be logically optimum if $est(C(\mathbf{q}) \mid e_n)$ were correct.[9]

From a Bayesian standpoint, the selection of a given prior Dirichlet - or, equivalently, the corresponding GC-system - implies obligation to update such a prior Dirichlet by Bayes's theorem. Such a commitment conflicts with the hyperempiricist solutions to EPO since they imply a completely different 'probability kinematics'. According to the hyperempiricist solutions, indeed, as new evidence e_n is gathered it is not used to *update* the selected prior Dirichlet by Bayes's theorem but to *replace* the prior Dirichlet with another prior (which, subsequently, is updated relative to e_n). This means that in the hyperempiricist approach probability kinematics is not governed by any prior Dirichlet.[10]

Hence, what initially appears as a solution to the problem of the *choice* of the optimum C-systems may be seen more properly as the *creation* of a new inductive system. Kuipers (*ibid.*, p. 44), for instance, seems aware that this is the case since he asks:

what kind of probability system results if I substitute for λ in [the formula for the special values of a λ-method] an estimate of λ^{A}? From the definition of a λ-method it is already clear that the result is not a λ-method, for an estimate of λ^{A} on the basis of e_n cannot be a *fixed* real number.[11]

9.3 THE PRESUPPOSITION VIEW

A clear formulation of the presupposition view of induction is given by Burks (1963, p. 758):

There is on the one hand, so to speak, the universe; it is this that we get information about, make predictions, adjust to, etc. There is, on the other hand, the [inductive method] which we use in learning about and adapting to the universe. Now the presupposition view about the relation of these two is roughly as follows. There are some quite general synthetic presuppositions about the universe which the use of the [inductive method] presupposes in the sense that if these presuppositions are in fact false the [inductive method] is not correctly applicable to the universe.

Although agreeing with Burks to some extent, I am inclined to think that certain 'local' presuppositions concerning *specific* universes (populations, processes) play a much more important role than the "quite general synthetic presuppositions about the universe" mentioned by Burks. In other words my approach to EPO emphasizes the 'local' and contextual character of the synthetic presuppositions used in selecting the epistemically optimum inductive method for investigating a given universe. For this reason my approach to EPO could be defined as a form of *contextual presuppositionism.*

The first systematic exposition of the 'presuppositionist view' of induction was given by John Stuart Mill who believed that the principle of uniformity of nature is an indispensable presupposition for induction. In his interpretation of Mill's views, Graves (1974, pp. 304-305) points out:

[Mill] certainly knows very well that many inductions do prove false, although a remarkable number succeed. In fact, he takes great pains to emphasize the diversity and complexity of nature as well its uniformity ... Furthermore, it is much too crude to say that nature either is or is not uniform, as if we had a simple dichotomy. What we have is a great many related particular uniformities, and the overall uniformity of nature is nothing more than the 'tissue' of these partial uniformities, or laws. We know from past experience that some uniformities do exist, and whenever we do argue inductively we are assuming (or believing, or hoping) that the case in question is one of them. We may of course be wrong; but to assume beforehand that we will be wrong is indeed inconsistent with the principle of uniformity.

On the basis of Graves's account of Mill's views, I would consider the contextual approach to EPO as a 'formal development' of Mill's idea that inductive procedures should be based on suitable assumptions about the "partial uniformities" of nature.

Graves (*ibid.*, p. 301) points out that "while Mill's own formulation [of the principle of uniformity of nature] is vague, it is not hopelessly so, and at least in special cases it can be made quite precise by using some of Carnap's ideas." More precisely, Graves (*ibid.*, pp.313-318) considers "Carnap's grandiose attempt to develop a fully quantitative inductive logic" and tries to "see how his ideas and Mill's might be used to clarify and supplement each other".

In particular, Graves (*ibid.*, p. 316) asks:

... what is the significance of λ ... and how are we to determine, estimate, or choose its value?

His answer (*ibid.*) is that:

Ontologically, λ provides ... a precise, quantitative, numerical scalar measure of the degree of uniformity of nature. The value we choose represents my estimate of how uniform nature is, at least in the area under investigation. We may make a poor choice, but the ontologically correct value determines the best rule to use. This value can be discovered only inductively, for it reflects my actual world and would be different in other possible worlds.

Afterwards, Graves (*ibid.*, p. 317) points out that:

... any justification of induction, or of a general inductive logic leading to specific rules of inference, must be based on non-logical facts about the constitution of the actual world and man's need to act in it. The world does have a degree of uniformity and if I knew just what it was in some *a priori* way I could use this to satisfy simultaneously the ideals of accuracy and efficiency in the best way possible. ... Since I do not know the true value, I still can and implicitly do make estimates. These are based largely on my previous experience of particular uniformities. ... Of course, even this picture is oversimplified. *I may find that nature is non-uniform in its uniformities, so that in different areas of investigation I may get better results by choosing different values of* λ. ...

... *in some form there is always an expectation, or at least a hope of some particular degree of uniformity.* [Author's italics]

Unfortunately Graves does not provide any concrete indication on how to estimate the "degree of uniformity" of the actual world. Therefore, he is unable to explain how to solve the related problem of 'estimating' the optimum value of λ.

It would appear that the presupposition view is also shared by Rosenkrantz (1981, Ch. 1.3, p. 1) who, after asking:

Is there any reason for preferring one λ-method to another? In particular, can we single out any of these methods as a uniquely reasonable inductive method?

suggests a negative answer to these questions (*ibid.*, Chapter 1.3., pp. 3-5):

... all that we can conclude is that ... methods with small λ are good in homogeneous universes, while ... methods with large λ perform well in heterogeneous universes. And, in practice, we do not know exactly what kind of universe we have by the tail.

... one's inclination is to conclude that *[the] rational choice of a λ-method should depend on how homogeneous one takes the population under consideration to be.* ... Yet, despite the fact that the optimal λ-method varies with the homogeneity of the considered population, Carnap continues to speak (1952, § 18) as if one chooses a λ-method for life ... But how we are to reconcile the claim that a λ-method is a long term policy to be applied across the board with the demonstrated dependence of the best method on the homogeneity of the considered population?[12] [Author's italics]

Also Hintikka stresses that certain synthetic assumptions play an important role in the choice of the optimum inductive method. For instance, referring to C-systems, Hintikka (1987, p. 303-304) states that "the optimal choice" of λ is determined by "the amount of order or disorder" in the investigated universe:

Hence, one's choice of the value of λ means guessing the relative degree of disorder in the world. The choice therefore cannot be made on logical grounds alone; it amounts to a synthetic assumption which nevertheless has to be made *a priori*. ... Indeed, my observation fits very well into the overall Bayesian strategy of building one's background information into the choice of prior probabilities. The use of parameters like λ may in suitable situations be an aid in so codifying one's background knowledge concerning the relative degree of order in the world.[13]

A presuppositionist view of TIP has also been advanced by Pietarinen (1972) - a follower of Hintikka - who holds (*ibid.*, p. 40) that a given inductive method can be considered rational only if it incorporates certain presuppositions about "the degree of uniformity of the phenomena in question".

Pietarinen (*ibid.*, p. 128) points out that "much of the old problem of justifying induction" is contained in the question about the rationality of the choice of a given inductive method. In particular, concerning the choice of λ, Pietarinen (*ibid.*, pp. 111-112) points out that:

... a rational choice of [the λ-value] ... accords with the presuppositions on the degree of 'regularity' of 'order' of the universe.

Having remarked that "the procedure of determining the values of the inductive parameters is *estimation*" (*ibid.*, p. 129) - in the sense that such determination is based on an estimation of the degree of order of a given universe - Pietarinen (*ibid.*, p. 131) recognizes that "it is difficult to spell out the exact principles that would yield an adequate method of estimation taking all relevant information into account" and, in fact, he does not suggest any specific procedure for estimating the degree of (dis)order in a given universe.

With reference to this, I wish to point out that the procedures illustrated in Chapter 8.2, for making external estimates of the degree of disorder $G(\mathbf{q})$ of a given population, represent a possible fulfillment of Pietarinen's request to specify the nature of the estimates which may be used as presuppositions in choosing a given inductive method.

I would also point out that the scientific applications of $G(\mathbf{q})$ and other diversity measures illustrated in Chapter 10 suggest that researchers, dealing with realistic problems in their area of expertise, *before* sampling a given population U under investigation can indeed produce meaningful 'informal estimates' of the Gini diversity $G(\mathbf{q})$ of U. Among other things, in Chapter 10 it is shown that $G(\mathbf{q})$ is equal to the physical probability that two individuals chosen at random and independently from U belong to different categories (see below formula (10.12)): hence an estimate of $G(\mathbf{q})$ amounts to an estimate of such a physical probability. Now it would appear plausible to assume that typically a scientist working in a certain area may informally estimate this probability even before sampling U. Lastly I would point out that - given the equalities $Est[G(\mathbf{q})] = E[G(\mathbf{q})]$ (see the requirement (CC), p. 98) and $E[G(\mathbf{q})] = 1 - \Sigma p(Q_i Q_i)$ (see (3.44.b)) - making an external estimate $Est[G(\mathbf{q})]$ amounts to establishing the value of the predictive probability that the next two individuals drawn at random and independently from U belong to different groups. Once again it seems that a researcher dealing with population or processes in his area of expertise can assess this predictive probability. In any way this assessment is not qualitatively different from - although usually more difficult than - the assessment of the predictive probability $\gamma_i = p(Q_i)$ that the next observed individual will belong to a given group Q_i.

9.4 THE VERISIMILITUDE VIEW

My V-solution to EPO is based on the assumption that researchers, in selecting an inductive method, attempt to maximize the possibility of approaching the truth. Such an assumption can be seen as a concrete application of a more general claim about the relationships between the scientific method and the truth made by Friedman (1978, p. 361):

> It would seem that some nice connection between confirmation and truth must exist if scientific method is to be justified as a rational activity. For one of the important ends of scientific method is the construction of theories that are true, or at least approximately true. Obviously, then, scientific method will be an effective means to this end only if it does indeed tend to produce true, or at least approximately true, theories.[14]

According to Friedman (*ibid.*, pp. 369) the requested connection between confirmation and the truth is guaranteed only by ***reliable*** inductive methods where the concept of reliability is informally elucidated thus:

> ... although a reliable method does not necessarily produce true conclusions, or even conclusion with a high probability of truth, it does *tend* toward truth: as time goes on, the conclusions that it accepts acquire ever increasing probabilities of truth. My suggestion is that it is the task of confirmation theory to show that the methods scientists actually use are reliable in this sense.[15]

From this concept of reliability it follows that there can *not* be any a priori justification for the scientific method. According to Friedman (*ibid.*, pp. 370-371)

> The impossibility of such a justification follows ... from two simple and fundamental facts: (i) there has to be some kind of link between justification and truth; a justification of scientific method must say something about its propensity to lead to truth; (ii) scientific method is not logically guaranteed of reaching true conclusions; it is an incurably nondeductive method. There is no inductive method that is more reliable in every logically possible world than every other method; consequently, there is no method that is a priori best, there is no method that is a priori the most reliable. We have to know facts about the actual world if we are to know which method is best; and we have to know facts about the actual world to know even that any given method has any chance at all of leading to truth.

Furthermore, Friedman (*ibid.*, p. 367) argues that the "superiority" of one inductive method over another is not an 'intrinsic feature' of that method; on the contrary, the fact that one of the two methods is a more effective tool for approaching the truth depends, to a large extent, on the structure of the external world:

How ... can ... one particular ... method be superior to the others? I think the answer is obvious: a method is superior if, when it is used in the inference I actually do make, it leads to true conclusions sooner and more often than the other methods. And which method actually enjoys this kind of superiority depends on empirical facts.

Consider, for instance, C-systems. Friedman (*ibid.*, p. 368, n. 3), having recalled that when λ is small (compared to n) the predictive probability of Q_i is closer to n_i/n, whereas when λ is large (compared to n) the predictive probability of Q_i is nearer to $1/k$, points out that:

Thus, the larger λ we choose, the more reluctant we are to extrapolate from my observed samples. Smaller values of λ work better in worlds in which observed samples (of almost any size) tend to be representative of the entire universe. Large values of λ work better in worlds in which observed samples (unless they are very big) tend not to be so representative.[16]

Note that, given theorem (8.22), it would appear quite natural to identify the "worlds in which observed samples ... tend to be representative of the entire universe" as the worlds with a low value of $G(\mathbf{q})$. If this assumption is correct, Friedman's assertion would imply that whenever a scientist expects $G(\mathbf{q})$ to be low - and he should accordingly expect the observed samples to tend to be representative of the entire universe - he should choose a small value of λ.

The importance of the concept of verisimilitude for comparing and selecting inductive methods is also stressed by Rosenkrantz (1980, p. 483):

The real importance of verisimilitude ... is that it constitutes a well-defined cognitive goal, leaving us free to compare inductive methods by the efficiency with which they realize this goal - that is by the expected truthlikeness of their outputs.

In particular, Rosenkrantz (*ibid.*, p. 485) claims that, from the concept of verisimilitude or truthlikeness:

... an interesting rationale can be given for adopting a diffuse prior where relevant data are lacking. For ... I can compare different priors by the expected truthlikeness of the posterior distributions to which they lead, conditional to the truth.

Rosenkrantz (*ibid.*, p. 485-87) explains the intuitive content of his proposal by referring to the case of "various beta priors for an unknown population proportion [$\mathbf{q}_1$]":

If your prior is heavily concentrated about the true value [of $\mathbf{q}_1$] (which amounts to a 'lucky guess' in the absence of pertinent data), you stand to be *slightly* closer to the truth after sampling than someone who adopts a diffuse prior, your advantage dissipating rapidly with

sample size. If, however, your initial estimate is in error, you will be farther from the truth after sampling, and if the error is substantial, you will be *much* farther from the truth. I can express this by saying that a diffuse prior is a better choice at 'almost all' values of [q_1] or, better, that it *semi-dominates* any highly peaked (or 'opinionated') prior. In practice, a diffuse prior never does much worse than a peaked one and 'generally' does much better. ...

It is well known that symmetric Beta priors, [Beta(α^*,α^*)] correspond to the 'methods' of Carnap's λ-continuum with [$\lambda = 2\alpha^*$] and that Carnap expressed a preference for small values of λ (i.e., for diffuse priors) without being able to justify that preference. He compared the λ-methods in terms of their mean-squared errors and showed that the optimal method is a function of the homogeneity of the considered population. It is not easy to reconcile this result with a preference for small values of λ ... to be used *across the board* in all estimation problems. A reconciliation can be effected, however, if we regard this preference as governing 'barren' contexts where no information about the population is at hand. In that case we can justify preference for diffuse priors (hence for estimates of the population proportion that are close to the observed sample proportion) by appeal to the average truthlikeness of the estimates to which they lead. The argument sketched here therefore fills a gap in Carnap's account.[17]

The assumption that verisimilitude provides the rationale for selecting the optimum inductive method is common to both my V-solution to EPO and that of Rosenkrantz. However, there is an important difference between the two approaches. Rosenkrantz's main objective is to justify the preference for "diffuse priors" (C-systems with a low λ) "in the absence of pertinent data".[18] On the contrary, I think that, in the analysis of scientific methodology, the most interesting inductive problems arise w.r.t. situations of partial information. As a matter of fact, my V-solution to EPO can be applied both to the cognitive contexts which include poor background knowledge and to those which include ample background knowledge.[19]

CHAPTER 10

DISORDERED UNIVERSES: DIVERSITY MEASURES IN STATISTICS AND THE EMPIRICAL SCIENCES

The contextual solution to EPO proposed in Chapter 8 can be applied in any G-context, i.e., in any multinomial context where an external estimate $Est[G(\mathbf{q})]$ of the *Gini diversity* $G(\mathbf{q}) = 1 - \Sigma \mathbf{q}_i^2$ of the population or process under investigation can be made (cf. Chapter 8.2). One of the principal aims of this chapter is to show that G-contexts, far from being merely theoretical possibilities, provide a realistic description of certain cognitive situations actually occurring in several empirical sciences. If this claim is true then the contextual solution to EPO can be concretely applied in several types of scientific research.

Among other things this Chapter should indicate that: (1) $G(\mathbf{q})$ is not an arbitrary or *ad hoc* index but provides a measure of the degree of disorder in a given population or process which appears to be useful and appropriate in several scientific disciplines; (2) in such disciplines $G(\mathbf{q})$ can be handled as any other 'respectable' scientific quantity and, in particular, the value of $G(\mathbf{q})$ in a given population or process can be conjecturally estimated before sampling on the basis of the background knowledge available to the researchers working in any given field. For instance, it seems clear that, as mentioned in Chapter 8.2, an estimate $Est[G(\mathbf{q})]$ of the Gini diversity of a given population could be made on the grounds of appropriate hypotheses concerning the relations between $G(\mathbf{q})$ and other characteristics of the population, such as stability, or its environment, such as pollution (see final page of Chapter).

Having supplied some historical remarks on Gini diversity (Section 1), several explications of diversity will be illustrated (Section 2) and, lastly, some scientific uses of Gini diversity and other diversity indices will be examined (Section 3).

10.1 GINI DIVERSITY

Pielou (1981, p. 408) defines a diversity index as

> **... a measure of the 'qualitative dispersion' of a population of individuals belonging to several qualitative different categories. In the same way that variance, standard deviation, mean, and range serve to measure quantitative variability, diversity indices measure qualitative variability.**

It would appear that the Italian statistician Corrado Gini (1912)[1] was the first to make a systematic analysis of qualitative variability. More specifically, he elaborated a unitary conceptual framework for the analysis of both quantitative and qualitative variability.

A key role in Gini's analysis is played by the concept of *mean difference* which is defined as follows.

Let the N individuals $a_1,...,a_N$ of a given population U be classified on the basis of a given quantitative characteristic - for instance weight - and let x_i denote the value of a_i, for $i = 1,...,N$. Then the mean difference $g(U)$ is defined as follows:

$$(1) \qquad g(U) = \frac{1}{N^2} \sum_{i,\, j=1}^{n} |x_i - x_j|$$

where the sum extends over all N^2 couples (x_i, x_j).[2]

The concept of mean difference may be adapted so that it may also be applied to various types of qualitative characteristics (see Gini, 1912, pp. 3-12). Consider, for instance, the family $\mathbf{Q}=(Q_1,...,Q_k)$ for which a 'trivial' similarity relation holds, in the sense that, for any four categories, or properties, $Q_i, Q_j, Q_h, Q_l \in \mathbf{Q}$ the degree of similarity between Q_i and Q_j is identical to that between Q_h and Q_l. Gini (*ibid.*, pp. 142-144) assumes that in this case the "*difference*" - or dissimilarity - between two individuals a_r and a_t in U is 0 if they have the same property Q_i, while it is 1 if they have two different properties Q_i and Q_j.

Let N_i be the number of Q_i-individuals in U. Then the sum of the differences between a given Q_i-individual a_r and each of the N members of U, including a_r, is given by: $N_i{\star}0 + (N-N_i){\star}1 = N - N_i$. Hence, the sum of all the N^2 differences between couples of members of U is given by $\Sigma N_i(N - N_i)$. Therefore the mean difference among the members of U is given by:

$$(2) \qquad G(U) = \frac{1}{N^2} \Sigma N_i(N - N_i)$$

Consider the vector $q=(q_1,...,q_k)$ where $q_i = N_i/N$. Given that $\Sigma N_i = N$, formula (2) can be rewritten as follows:

$$(3) \qquad G(U) = G(q) = 1 - \Sigma q_i^2$$

which is the usual definition of $G(q)$.

For a long period Gini's research on qualitative variability was little known outside Italy and even today the historical importance of Gini's work on the heterogeneity of populations has still to be fully recognized.[3] As a consequence of the relative lack of knowledge of Gini's work on this subject, the concept of mean difference has been 'rediscovered' by several statisticians and scientists. For instance, according to Patil and Taillie (1982, p. 556) the diversity of a given community can be defined as "the average differentness [i.e., mean difference] between two randomly selected members of the community".

Analogous to the mean difference explications of diversity proposed by Gini and other authors, mean similarity explications of uniformity have been advanced. For instance, given a measure of similarity S "between any pair of animals (or more in general units)" in a given population, the mean similarity $E(S)$ among the units in the population could be considered as the "*weighted* index of homogeneity" or the "*index of uniformity*" of the population (Good, 1982, p. 562). Note that if S is a trivial similarity function which can only assume the values 0 and 1 then $E(S)$ is equal to Σq_i^2 (see Good, *ibid.*). This means that Carnap's uniformity index $C(q) = \Sigma q_i^2$ can be seen as the result of applying Good's mean similarity explication of uniformity to a 'trivial-similarity' family $\mathbf{Q}=(Q_1,...,Q_k)$.[4]

10.2 EXPLICATING DIVERSITY

Average rarity explications

Average rarity explications of diversity are based on "the view that diversity is an average property of a community" where such a property is given by "species rarity" (Patil and Taillie, 1982, p. 548). Let $Rar(Q_i,U)$ be a measure of the *rarity* of the species Q_i in a given community U. Then, the average rarity explicatum $Div(U)$ can be defined as follows:

$$(4) \qquad Div(U) = \Sigma\, q_i\, Rar(Q_i,U)$$

Different diversity indices can be obtained from (4) depending on the selected rarity measure $Rar(Q_i, U)$.

Consider a "dichotomous-type rarity measure" $Rar(Q_i, U) = Rar(q_i)$ which depends only on the "relative abundance" q_i of Q_i in U (Patil and Taillie, 1979, p. 6).[5] Then the intuitive requirement that rarer species correspond to smaller values of q_i can be expressed as follows:

(RAR) $Rar(q_i)$ is a decreasing function of q_i.[6]

The requirement (RAR) is satisfied, for instance, by the following definition of species rarity:

$$(5) \qquad Rar(q_i) = 1 - q_i$$

If the average rarity explicatum (4) and the definition (5) of species rarity are adopted then Gini diversity is immediately obtained: $Div(U) = \Sigma q_i(1 - q_i) = 1 - \Sigma q_i^2 = G(q)$.

Similarly some homogeneity indices have been interpreted as measures of the 'average commonness' of the different species present in a given population. Let $Com(Q_i, U)$ be a measure of the *commonness* of the species Q_i in a given community U. Then the degree of homogeneity $Hom(U)$ can be defined as:

$$(6) \qquad Hom(U) = \Sigma\, q_i Com(Q_i, U)$$

Note that $C(q)$ can be interpreted as a measure of average commonness, provided that $Com(Q_i, U)$ is equated to the relative frequency q_i of Q_i in U:[7]

$$(7) \qquad Hom(U) = \Sigma\, q_i \cdot q_i = \Sigma\, q_i^2 = C(q)$$

Hill (1973, p. 428) introduces the following parametric family of diversity indices based on the concept of "proportional abundance", or commonness, of the different species in population U:

$$(8) \qquad N_a(q) = \left[\frac{w_1 q_1^{a-1} + w_1 q_2^{a-1} + \ldots + w_k q_k^{a-1}}{w_1 + w_2 + \ldots + w_k} \right]^{1/(1-a)}$$

where $N_a(q)$ is called the "diversity number of order a" (*ibid.*)

Hill (*ibid.*) points out that $N_a(q)$ "is the reciprocal of the $(a\text{-}1)^{\text{th}}$ root of a weighted mean of the $(a\text{-}1)^{\text{th}}$ powers of the proportional abundances" of the k species $Q_1,\dots,Q_k$ present in U. Therefore $N_a(q)$-diversities can be "characterized as reciprocals of mean proportional abundances".

Assuming that each of the weights w_i occurring in (8) is equal to q_i, then (8) can be rewritten as follows:

$$(9) \qquad N_a(q) = (q_1^a + q_2^a \dots + q_k^a)^{1/(1-a)} = (\Sigma\, q_i^a)^{1/(1-a)}$$

The family $N_a(q)$ and the related family of uniformity measures $1/N_a(q)$ include several well known indices. For instance, it follows from (9) that $N_2(q)$ is equal to $1/(\Sigma q_i^2)$; hence the uniformity measure $1/N_2(q)$ is equal to Carnap's homogeneity index Σq_i^2.[8]

Distance-from-maximal-heterogeneity explications

A possible explication of the notion of homogeneity is based on the following two step strategy (cf. Bhargava and Doyle, 1974, p. 243):

(i) identify the 'even' vector q^* which represents the case of *maximal heterogeneity*,

(ii) define the homogeneity $Hom(q)$ of a given population U with vector q as the distance between q and q^*:

$$(10) \qquad Hom(q) = D(q,q^*)$$

If the quadratic distance $D_2(q,q^*)$ is applied and q^* is equated to the indifference vector $\mathbf{1/k}=(1/k,\dots,1/k)$,[9] then the homogeneity index derived from (10) is a linear function of Carnap's index $C(q)$:

$$(11) \qquad Hom(q) = D_2(q,\mathbf{1/k}) = C(q) - 1/k.$$[10]

A similar strategy could be used for explicating diversity. Indeed, given an appropriate measure of the 'proximity' between two probability vectors, the diversity $Div(q)$ of U could be defined as the proximity between q and q^*.[11]

Probabilistic and encounter-theoretic explications

The probabilistic explications of homogeneity and heterogeneity date back to at least to Simpson (1949, p. 688) whose "measure $[\Sigma q_i^2]$ of the concentration of the classification" of a given population U in groups derived from his preceding collaboration with Turing (cf. n. 4).

According to Simpson (*ibid.*, p. 688), indeed, the homogeneity index Σq_i^2 "can be simply interpreted as the probability that two individuals chosen at random and independently from the population will be found to belong to the same group", as emerges from the equality:

$$(12) \qquad C(q) = \Sigma q_i^2 = \Sigma q_i \cdot q_i$$

Similarly, the Gini diversity $1 - \Sigma q_i^2$ of U is equal to the probability $\Sigma q_i(1-q_i)$ that two individuals chosen at random and independently from U belong to different groups.[12]

More recently, Good (1982, p. 562) proposed the following parametric family of homogeneity indices:

$$(13) \qquad C_s(q) = \Sigma q_i^s \qquad \text{where } s \geq 2$$

where $C_s(q)$ is the probability that s units, randomly selected with replacement from a given population, will belong to the same category. Note that Carnap's index is included in Good's family: $C(q) = \Sigma q_i^2 = C_2(q)$.

In some scientific disciplines an *encounter-theoretic approach* to diversity is proposed which provides a kind of 'operational interpretation' of the probabilistic explications described above. While such explications make reference to the probabilities of 'artificial events' such as the random selection of two or more individuals from a given population, the encounter-theoretic approach to diversity makes explicit reference to the probabilities of 'natural events' such as the spontaneous intraspecific and interspecific encounters between couples of individuals within a certain community.[13]

For instance, in the biology of populations, the encounter-theoretic concept of diversity has been considered by Hurlbert (1971, p. 577) who argues that many commonly used diversity indices lack any evident "empirical interpretation" and suggests an alternative set of indices having "straightforward biological interpretations".

In particular, the interpretation advanced by Hurlbert makes reference to "the probability of interspecific encounters (PIE)" (*ibid.*, p. 579). The intuitive meaning of PIE is illustrated by Alfred Russell Wallace's vivid description of an Amazonian forest (1876, p. 75: quoted by Hurlbert, *ibid.*):

If the traveller notices a particular species and wishes to find more like it, he may turn his eyes in vain in any direction. Trees of varied forms, dimensions and colors are around him, but he rarely sees any one of them repeated. Time after time he goes towards a tree which looks like the one he seeks, but a closer examination proves it to be distinct. He may at

length, perhaps, meet with a second specimen half a mile off, or may fail altogether, till on another occasion he stumbles on one by accident.

Wallace's traveller may be seen as an intruder - for instance a biologist or an other organism - entering a community and randomly encountering individuals belonging to it.

Referring to a community with N individuals and k species $Q_1,...,Q_k$ - where $q_i = N_i/N$ is the relative frequency, or "abundance", of Q_i - Hurlbert (*ibid.*) points out that "when the first individual encountered risks being the subject of the second encounter also, as in non lethal encounters" the probability that two individuals encountered by the intruder belong to different species is simply $\Delta^2 = 1 - \Sigma q_i^2$. Notice that Hurlbert's index Δ^2 is the same as Gini diversity.[14]

Hurlbert (1971, p. 581) also considers the following parametric family of diversity indices $E(S_n')$:

(14) $$E(S_n') = \Sigma\,[1 - (1 - q_i)^n]$$

where $E(S_n')$ is the number of species which, on average, would be met by an individual "which enters a community and in a certain period of time encounters n individuals at random", provided that "the intruder eats none of the individuals encountered" (Hurlbert, *ibid.*).

It can easily be proved that the member $E(S_2')$ of the family $E(S_n')$ is linked to $G(q)$ by the following relation:

(15) $$E(S_2') = 1 + G(q)$$

Requirements of adequacy for diversity

Whenever the members of a given population U are classified w.r.t. a 'trivial-similarity' family $\mathbf{Q}=(Q_1,...,Q_k)$, it would seem intuitively plausible that at least the following 'minimal' requirements of adequacy are satisfied by a diversity measure $Div(U)$ for U:

Requirement of q-sufficiency

$Div(U)$ is a function, at most, of $q=(q_1,...,q_k)$. Hence, $Div(U)$ can be referred to as $Div(q)$.

Requirement of symmetry

$Div(q)$ is a symmetrical function of q, i.e., if π is a permutation of $(q_1,...,q_k)$ and $q'=(\pi(q_1),...,\pi(q_k))$, then $Div(q') = Div(q)$.

Requirement of maximization
$Div(q)$ reaches its maximum value when q is identical to the indifference vector $\mathbf{1/k}=(1/k,...,1/k)$.

Requirement of minimization
$Div(q)$ reaches its minimum value when q is one of the k 'maximally homogeneous' vectors: (1,0,...,0), (0,1,0,...,0), ... , (0,...,0,1).

Requirement of evenness
$Div(q)$ increases when two components q_i and q_j of q, are 'flattened' towards $1/k$ in the following sense. Let $q=(q_1,...,q_k)$ and $q'=(q'_1,...,q'_k)$ be two vectors so that $q_i < q'_i \leq 1/k \leq q'_j < q_j$ and $q'_h = q_h$ for all $h \neq i,j$. Then $Div(q') > Div(q)$.

Indeed, these minimal requirements - which may be seen as the core of presystematic intuitions concerning diversity - appear to be satisfied by all the diversity measures mentioned in this Section, including Gini diversity.

However, they do not seem sufficient for the axiomatic derivation of Gini diversity or other diversity measures. For instance, the few available axiomatic characterizations of Gini diversity (cf. Bhargava and Uppuluri, 1975; Rao, 1982) make use - in addition to some of the above minimal requirements - of other and stronger principles (where the intuitive meaning and plausibility of some of these principles are not completely clear).

Several authors after Gini (1912) pointed out a number of 'plausible properties' of diversity measures which can be interpreted as more or less explicit (minimal) requirements of adequacy for diversity (see, for instance, Agresti and Agresti, 1977, pp. 206-207). It may be asked whether these requirements could be used to explicate diversity on the basis of the following two step approach widely adopted in the analytical philosophy to explicate methodological concepts:

(i) presystematic intuitions about the diversity of a population are formally expressed by an appropriate set of requirements of adequacy;
(ii) the most simple explicatum satisfying such requirements is adopted.

Of course this approach does not exclude the following two possibilities: (i) the adopted explicatum is given by a *continuum* of diversity measures; (ii) not only is such a continuum compatible with the requirements of adequacy but it is *axiomatically derivable* from them.

Although this approach - as opposed to the above described explications of diversity derived from a single 'monolithic' intuition - has not yet been systematically investigated, I believe it merits serious consideration.

10.3 DIVERSITY MEASURES IN THE EMPIRICAL SCIENCES

According to Sugihara (1984, p. 564), the notion of diversity has both a philosophical and a scientific interest:

> There are two broad reasons why attention has been focused on characterizing the diversity of an ecological assemblage. These depend on whether one treats diversity as an abstract phenomenological property or a specific biological one. That is, whether one treats diversity in its general sense, as a property in itself, or as an indicator of the functioning and organization of communities. First, diversity per se, as a phenomenological property, contains intrinsic interest, and as a fundamental quality of perception seems to demand quantification. It is not surprising, therefore, that one finds diversity measures emerging in such varied disciplines as genetics, linguistics, and economics. ...
>
> Here justification for a particular index rests more on philosophy than science, since the primary purpose of the index is to reflect a human value rather than to capture an important property of state in the functioning of communities.

As seen above several diversity measures may be obtained as the result of 'philosophical explication' of the presystematic concept of diversity. A purely theoretical analysis of diversity may also be useful in the empirical sciences since it reveals the 'logical structure' of diversity measures.

However, the choice of a given diversity measure $Div(q)$ is entirely up to the scientist. Indeed, the scientific usefulness of $Div(q)$ depends on the discovery of certain relations between the diversity $Div(q)$ of a population and further characteristics of the population or of other entities related to it.[15] In other words, the scientific usefulness of $Div(q)$ depends on the scientists' ability to link $Div(q)$ with other scientific quantities in a network of well confirmed empirical laws.[16]

More specifically, Sugihara (*ibid.*, p. 564), referring to diversity as a biological property, remarks that:

> To the ecologist, diversity is interesting as a property of state in so far as it has the potential to reflect the nature of the underlying processes and organization that structure the community. Therefore, beyond arbitrary or weakly motivated definitions, *the scientific interest in and importance of ecological diversity hinges directly on its possible connections with the functioning and organization of communities*. The principal aim of such scientific study, therefore, is to find a characterization of species diversity that most clearly reveals this functional connection.[17] [Author's italics]

With reference to this, remember that the encounter-theoretic interpretations of Gini diversity and other indices (see Section 2) were forwarded in an attempt to identify diversity measures potentially useful for understanding the "functioning and organization of communities" where *interactions* between members of different categories visibly play an important role in the life of the community.[18]

Concerning the *scientific applications of Gini diversity G(q)* and other diversity measures in the empirical sciences, it should be remembered that $G(q)$ is applied in several ***human sciences*** such as sociology (Lieberson, 1969), economics (Amemiya, 1963) and linguistics (Yule, 1944; Greenberg, 1956; Guiraud, 1959; Herdan, 1964, 1966).[19]

$G(q)$ and other diversity and uniformity measures have also been used in several ***natural sciences*** such as genetics (Lewontin, 1972), biology (Gatlin, 1972) and ecology (see, for instance, Pielou, 1969, 1975, 1977 and Grassle, Patil, Smith and Taillie, 1979).

Some interesting comments on the 'intrinsic' importance of the ecological concept of diversity are made by Dennis, Patil, Rossi, Stehman and Taillie (1979), in the introduction to an exhaustive bibliography - including 1046 titles - on the literature on ecological diversity:

> "Why diversity?" Recently a statistician noticed that he never received a simple straightforward answer to this question from ecologists. But imagine the look of disappointment in an ecologist hearing this question. The diversity, or variety, of plants and animals on this planet is the very basis of the ecologist's profession.
>
> Ecology is rooted in natural history. Perhaps the fundamental working craft of ecologist is taxonomy; possibly the most elemental quantities recorded in ecological work are the numbers and abundances of species. Why are there so many species in the abundances that we observe? Indeed, almost any ecological study has some bearing on this question, as these quantities will be explained only after the interrelationships of the organisms and environments are fully understood.
>
> So if you ask an ecologist "why diversity?" the response, if any, will likely echo the retort of the mountaineer Mallory when questioned about his own pursuits: "Because it is there!".

This quotation clearly illustrates that an intrinsically interesting 'phenomenological parameter' of an ecosystem is its degree of diversity: as such, diversity should be the *explanandum* of appropriate scientific explanations.

On the other hand, a certain diversity measure could also be used as an *explanans* to explain, or predict, other features of a given environment. With reference to this, Patil and Taillie (1982, p. 566) remark that:

> Research on diversity as an input parameter is relatively in its embryonic stage. A major question is, What of consequence, if any, does the diversity of a system help predict or determine? Some of the questions that have been pursued are, Does the diversity of a

community help predict its stability? Does the biotic diversity of a habitat help predicting its abiotic classification? Does the geological diversity of a region help predict its mineral diversity?[20]

The importance of the relationships between diversity and other scientifically interesting parameters is also pointed out by Pielou (1975, p. 6):

The purpose of measuring a community diversity is usually to judge its relationship either with other community properties such as productivity and stability, or to the environmental conditions that the community is exposed to.

With reference to this, it should be pointed out that over the last few years there has been an increasing theoretical and practical concern with the possible impact of environmental pollution on the biological diversity of ecosystems[21] and, consequently, the possible relationships between pollution and diversity have been extensively investigated.[22]

CHAPTER 11

CONCLUDING REMARKS

In this chapter the relationships between the contextual view of prior probabilities and some recent methodological programmes of research are considered; moreover some possible developments of the contextual view are suggested.

Relationships between the contextual view of prior probabilities and other methodological programmes of research

In spite of the increasing attention paid to scientific revolutions over the last thirty years, it is generally recognized that most researchers are concerned with what Kuhn calls *normal science*. Therefore the methodology of normal science continues to be of great importance in epistemology. Moreover, the analysis of the descriptive and inductive methods of normal science appears to be a primary goal of statistics. A number of recent studies on normal science attempt to combine some conceptual tools borrowed from epistemology with others borrowed from statistics (see Rosenkrantz (1977, 1981), Good (1983) and Howson and Urbach (1989)). The contextual approach to prior probabilities is part of this methodological tendency since it suitably integrates some conceptual tools borrowed from epistemological studies on the optimum inductive methods and others borrowed from statistical studies on the choice of prior distributions.

Prior probabilities denote the initial probabilities which scientists attribute to the hypotheses under consideration before performing a given experiment (cf. Chapter 3.1). The subjective view and the aprioristic view represent the 'traditional' standpoints on the choice of prior probabilities. According to the subjective view the choice of prior probabilities is restricted only by the probability axioms. Since such axioms are compatible with an extraordinary variety of prior distributions subjectivists think that there cannot be any rational consensus among scientists before performing the experiment (cf. Chapter 7.1 and 7.3). According to the aprioristic view, on the contrary, the choice of prior probabilities is restricted not only by the probability axioms but also by certain a priori principles which always allow identification of a unique admissible prior distribution; hence there cannot be any rational disagreement among scientists before carrying out the experiment (cf. Chapter 7.2 and 7.4). The choice of prior probabilities is one aspect of the more general issue concerning the rational (dis)agreement in science. While the subjective view of prior probabilities explains the possibility of rational disagreement and the aprioristic view that of rational consensus, neither view succeeds in explaining how *both* rational disagreement and rational consensus can occur before performing the experiment.[1]

As pointed out by Laudan (1984, p. 3), an important task of methodological analysis is to delineate "some machinery that explains how both consensus and dissensus can arise". It seems to me that the contextual view of prior probabilities helps to tackle this task. Indeed according to this view the choice of prior probabilities is restricted not only by the probability axioms but also by certain contextual principles and other 'contextual constraints' which depend on the cognitive context of the researcher (cf. Chapter 7.5). This implies that if the cognitive contexts of several researchers are identical, there should be complete rational consensus among them, while if such cognitive contexts are to some extent dissimilar, a corresponding degree of rational disagreement may arise.

According to an epistemological view called *local induction* (cf. Bogdan, 1976) the appropriate rule of inductive acceptance for a given inquiry can be seen as the 'output' of a number of different 'contextual inputs' (such as the problem under consideration and the cognitive goals of the inquiry). Although local induction has so far dealt almost exclusively with the rules of inductive acceptance, it might more generally be defined as a methodological programme of research which aims to provide a contextual justification for the inductive procedures used in scientific research. The contextual approach to prior probabilities can be seen as a possible development of local induction. Indeed the optimum prior distribution for a given inquiry is seen as the output of certain contextual inputs such as the available background knowledge and the cognitive goals of the inquiry (cf. Chapters 7.5 and 8).

The verisimilitude theory might be seen as a particular formulation of local induction. In other words VT can be seen as a methodological programme of research which aims to justify inductive procedures w.r.t. to a specific cognitive goal represented by the achievement of a high degree of verisimilitude. The 'verisimilitude programme' can deal with a wide variety of cognitive contexts since the verisimilitude thesis (VER) underlying VT leaves considerable freedom to select the measure of verisimilitude (distance from the truth) to be adopted in a given inquiry.[2] The V-solution to EPO proposed in Chapter 8.5 can be seen as a contribution to the verisimilitude programme since the optimum prior Dirichlet distribution Dir($\gamma°$,$\lambda°$) is intended to minimize the distance between certain Dir($\gamma°$,$\lambda°$)-based point estimates of the investigated parameter vector and the truth.

Some possible developments of the contextual view

In Chapter 7.5 a contextual justification of the requirement of neutrality (CM) and, accordingly, of the decision to use a Dirichlet prior (a GC-system) was given. Some possible developments of the contextual view of prior probabilities are: (i) to provide a contextual justification for appropriate families of prior distributions (inductive methods) for various kinds of multinomial contexts where (CM) does not hold;[3] (ii) to provide a contextual justification for appropriate families of prior distributions (inductive methods) for Markov processes and other multicategorical processes different from multivariate Bernoulli processes;[4] (iii) to provide a contextual justification for appropriate families of prior distributions for various types of 'quantitative processes' such as the normal process and the Poisson process.[5]

The CC-solution to EPO for a given set of prior Dirichlet proposed in Chapter 8.3 was based on an external estimate of the Gini diversity of the investigated multivariate Bernoulli process. However 'alternative CC-solutions' to EPO based on external estimates of other diversity measures deserve to be studied.[6] The problem of identification of the epistemically optimum prior for a given inquiry (EPO) may be defined not only w.r.t. a given set of Dirichlet priors but also w.r.t. other sets of priors applicable to various kinds of quantitative processes. The possibility of using external estimates of the 'quantitative variability' of the considered processes to work out adequate CC-solutions to such EPO's deserves careful examination.[7]

The V-solution to LPO for a given set of Dirichlet priors was defined w.r.t. *M*-estimators and the quadratic distance used in the verisimilitude interpretation of such estimators (see Chapter 8.4). Moreover the V-solution to LPO was used - in combination with (VER) - to work out a V-solution to EPO (see Chapter 8.5). It seems to me that two important problems deserve to be carefully examined: (i) the possibility to work out alternative V-solutions to LPO, defined w.r.t. other Bayesian (point or interval) estimators and other distance functions,[8] (ii) the possibility to use such V-solutions to LPO for developing corresponding 'alternative V-solutions' to EPO.[9]

Remember that a V-solution to LPO can be defined not only w.r.t. a given set of prior Dirichlet but, more generally, w.r.t. any set of prior distributions on the parameter (vector) of a given experimental process (see Chapter 8.4).

It seems to me that a reason for carefully exploring certain V-solutions to such LPO's is to ascertain whether they can help work out adequate V-solutions to the corresponding EPO's.

NOTES

CHAPTER 1

[1] After World War II the Bayesian approach has gained increasing support among statisticians and epistemologists. In particular, in the last fifteen years the Bayesian approach to epistemology has been developed by several authors such as Rosenkrantz (1977,1981), Horwich (1982), Levi (1980), Howson and Urbach (1989). Although the contention between the Bayesian approach and other statistical and epistemological approaches is an intriguing topic (see, for instance, Barnett (1973), Howson and Urbach (1989), Earman (1992)) it is not my topic since my concern in this book is with problems arising *within* the Bayesian approach.

[2] The theory of inductive probabilities (developed by Carnap and other epistemologists) deals with certain types of inductive inferences, such as prediction of future events, which are also typical subjects in philosophical research on induction. In particular, the problem of assessing the probability of a future event - which had already been considered by Hobbes (1650) - has received much attention after it was studied by Hume (1739): see Hacking (1975, p. 48 and 178).

[3] Following a common epistemological usage, here "hypothesis" refers to whatever factual statement, from specific predictions to highly general theories. On the contrary, in statistics "hypothesis" is typically used with a much narrower meaning. Unfortunately, many other terminological conflicts occur between epistemology and statistics. Although I tried to make the text sufficiently clear to readers with a background in any of the two fields, a bias towards the epistemological jargon was inevitable given *my* own background. Hence, some tolerance is requested to those readers who will find familiar terms employed with a unfamiliar meaning (or vice versa).

[4] This term is used, among others, by Swinburne (1973) and Skyrms (1966).

[5] Of course scientists *qua* human beings pursue non-cognitive goals even *within* their scientific activity. For instance, "many people called scientists regard science as some battlefield where 'being regarded as clever and correct' is more important than 'really having done all the work'" (Prof. W. Schaafsma, private communication).

[6] The notion of cognitive context is borrowed from Levi (1967).

[7] Empiricist philosophers such as Francis Bacon and rationalist philosophers such as Descartes, although holding different views about the *nature* of scientific method, shared the same infallibilistic view - which Watkins (1978, p. 25) calls the "Bacon-Descartes ideal" - about the *goal* of science. The infallibilistic ideal was also advocated by scientists and philosophers such as Boyle, Locke and Newton.

[8] Here "the truth" is intended in the sense of the *correspondence theory of truth* suggested by Aristotle and accepted, more or less explicitly, by most infallibilists: a statement is true if and only if it *corresponds* with reality, i.e., with the way things really are. According to the infallibilistic view the scientific method, if properly used, infallibly guarantees the discovery of true theories (see Laudan, 1973, p. 277).

[9] Indeed certainty about a given statement may be seen as the maximum degree of belief in the truth of the statement.

[10] For instance, Descartes (ca. 1628) maintains that we should "reject ... merely probable knowledge and make it a rule to trust only what is completely known and incapable to be doubted."

[11] A fascinating inquiry into the 'emergence of probability' in modern thought is made by Hacking (1975).

[12] The origins of the concept of verisimilitude and, more generally, the fallibilistic methodologies are traced by Niiniluoto (1987, Ch. 5).

[13] It would appear that in the last century a number of philosophers had already recognized that the probabilistic and verisimilitude views are not incompatible (see Laudan, 1973, pp. 285-286 and 295). However, according to Laudan (*ibid.*, p. 286) "these two approaches did ... represent different emphases, and were to give rise in the twentieth century to two very different strains in philosophy of science (Carnap and Keynes being the descendants of the progress by probabilification school, and Popper and Reichenbach focusing primarily on progress by self-correction)".

[14] Another example of a co-ordinated development of the Bayesian approach and the verisimilitude theory is given by the notions of *expected verisimilitude* and *probable verisimilitude* which can be defined using the concepts of epistemic probability and verisimilitude (see Chapter 4.3).

[15] Following a common epistemological usage, here "inference" is employed in the sense "argument". This conflicts with the more frequent statistical usage of "inference" in the sense of "conclusion" (cf. note 3).

[16] The conclusion of an inductive inference, indeed, is frequently termed "hypothesis" because of its conjectural character.

[17] As pointed out by Hacking (1975, Chapter 2), since its emergence in the Western thought probability was essentially dual, on the one hand having to do with degrees of belief

(*epistemic probability*), on the other, with devices tending to produce stable long-run frequencies (*physical* or *objective probability*).

[18] See Chapter 2.1 and Chapter 11, note 4.

[19] The term "Generalized Carnapian (GC-)systems" is borrowed from Kuipers (1978).

[20] See Chapter 7, note 18.

[21] Gini diversity is a measure of the *degree of disorder* of a population or process (for a more detailed description see Chapter 10). Contrary to what one might believe at first sight, $Est[G(\mathbf{q})]$ is different from Gini diversity $G(\gamma^\circ)$ of the prior vector γ°: see formula (8.7) and Chapter 8, note 5.

[22] On the other hand idealizations - which play a basic role in empirical sciences - may be very useful also in methodological analysis. This basically depends on the possibility of finding interesting generalizations and 'concretizations' of the proposed idealizations (cf. Chapter 11 where some possible extensions of the present approach to EPO are suggested).

CHAPTER 2

[1] All non-quantified variables - such as n, e_n and e'_n - contained in the formulae and requirements of adequacy stated herein should be interpreted as universally quantified.

[2] The axiom of convergence was suggested to Carnap in 1953 by Hilary Putnam, who proposed for it the name "Reichenbach axiom" (cf. Carnap, 1980, p. 120).

[3] It can also be proved that (Exc) is logically equivalent to the conjunction of two requirements given in terms of special values (see Kuipers, 1978, pp. 40-41).

[4] Later this view was defended by Carnap (1950, pp. 208 and 571-575).

[5] Apart from some slight terminological modification, this classification is borrowed from Carnap (1950, pp. 205-208).

[6] The inference from a sample to the population is usually referred to as the *inverse inference* (cf. Carnap, 1950, pp. 207-208).

[7] On the whole, Carnap devotes little attention to statistical inferences. He only provides some formulae to calculate the probability of point hypotheses such as "$\mathbf{q}_1 = 0.3$" for a *finite* population (Carnap, 1950, p. 570 and 1952, pp. 30-32). As far as I know during the fifties one of the few attempts to investigate the possible statistical implications of Carnap's TIP was made by Tintner (1949) who referred to the manuscript of *The Logical Foundations of Probability* (Carnap, 1950).

[8] For a more detailed description of (multivariate) Bernoulli processes see Feller (1968, Ch. 6) and La Valle (1970, Ch. 11).

[9] The notions of "multivariate Bernoulli process/multinomial context" used herein correspond approximately to the notions of "repeatable experiment/multinomial context" as defined by Kuipers (1978, pp. 114-115).

CHAPTER 3

[1] Various *rules* of inductive acceptance have been proposed in statistics and epistemology. Furthermore, the *meaning* of the inductive acceptance of a hypothesis has been interpreted in different ways. While the 'weakest' type of acceptance presumably involves selection of a hypothesis for further scrutiny (see, for instance, Popper's methodological rules and Fisher's significance tests), the 'strongest' type presumably involves the incorporation of the accepted hypothesis within the available background knowledge which will be taken as guaranteed in subsequent inquiries (see, for instance, Levi's acceptance rules (1976)). Other rules of inductive acceptance seem to lie somewhere between these two 'extremes'. For instance, point estimation could be seen as a type of 'as if-acceptance' where the estimated value of a given parameter is tentatively used, especially in calculations. A stronger type of inductive acceptance appears to be represented by interval estimation, and an even stronger type by the Neyman-Pearson hypothesis testing.

[2] The probabilistic approach has been supported by Carnap (1968a), Jeffrey (1956, 1968, 1970), Bar-Hillel (1968), Box and Tiao (1973, Appendix A5.6) and others. The supporters of the acceptational approach include, among others, Hempel (1962), Hintikka and Pietarinen (1966), Hilpinen (1968), Kyburg (1974), Levi (1967, 1976, 1980), Kaplan (1981), Harsanyi (1985), La Valle (1970) and Zellner (1971).

[3] The origins of the predictivistic approach to statistical inference date back at least to 1774 with the famous Laplace's rule of succession (cf. Zabell, 1989). The predictivistic approach received strong backing from Roberts (1965). A systematic exposition of the predictive methods used in statistics is given by Aitchison and Dunsmore (1975).

In conflict with the *monistic approaches* mentioned in the text - which emphasize the methodological relevance of *one* of the different kinds of inferences - it could be argued that different kinds of inductive inferences are appropriate in different situations. For instance, it seems clear that predictive inferences play a crucial role in risk analysis while in theoretical physics global inferences are much more important. This suggests that a *pluralistic approach* would provide a more adequate image of the inductive inferences demanded in scientific practice.

[4] The predictive distribution $p(y/z)$ specifies the epistemic probabilities of the possible results of a *future* experiment **y** relative to the result z of a *previously performed* experiment **z**.

[5] As Winkler (1972, p. 143) points out, continuous distributions "are studied because (1) they are often mathematically easier to work than discrete models, and (2) they provide an excellent approximation to numerous discrete models."

[6] In the first case the probability centres on single values of the parameter vector and can be based on a *joint probability function* $p(\theta_1,\theta_2)$ analogous to the one described in the univariate case. In the second case it covers the entire plane or some subregion of it and no single point receives a positive probability.

[7] Here $p(\theta/y)$ and $p(\theta)$ are either both probability functions or both density functions, according to whether $\boldsymbol{\theta}$ is discrete or continuous, and $p(y/\theta)$ and $p(y)$ are likewise either both probability functions or density functions according to whether **y** is discrete or continuous.

[8] Cf. Aitchison and Dunsmore (1975, pp. 1 and 17-23).

[9] Formula (36) can be proved as follows.

From (22)(b) and the equalities $E(\mathbf{q}_i\mathbf{q}_j) = p(Q_iQ_j)$, $E(\mathbf{q}_i) = p(Q_i)$ and $E(\mathbf{q}_j) = p(Q_j)$ (see (34) and (32)), it follows that $\text{cov}(\mathbf{q}_i,\mathbf{q}_j) = p(Q_iQ_j) - p(Q_i)p(Q_j) = p(Q_i)p(Q_j/Q_i) - p(Q_i)p(Q_j) = p(Q_i)[p(Q_j/Q_i) - p(Q_j)]$, *Q.E.D.*

[10] Formula (40) follows from $\text{var}(\mathbf{q}_i\,|\,e) = p(Q_iQ_i/e) - [p(Q_i/e)]^2$ (see (35)), $E(\mathbf{q}_j\,|\,e) = p(Q_j/e)$ and the sequence of equalities: $p(Q_iQ_i/e) - [p(Q_i/e)]^2 = p(Q_i/e)p(Q_i/eQ_i) - p(Q_i/e)p(Q_i/e) = p(Q_i/e)[p(Q_i/eQ_i) - p(Q_i/e)]$.

[11] Formula (41) follows from the equalities $E(\mathbf{q}_i\,|\,e) = p(Q_i/e)$ and $\text{cov}(\mathbf{q}_i,\mathbf{q}_j\,|\,e) = p(Q_i/e)[p(Q_j/Eq_i) - p(Q_j/e)]$ (cf. (36)).

[12] Formula (42) follows from the sequence of equalities: $\text{VAR}(\mathbf{q}) = \Sigma\text{var}(\mathbf{q}_i)$ (from (37)) $= \Sigma(E(\mathbf{q}_i^2) - [E(\mathbf{q}_i)]^2)$ (from (20)(b)) $= \Sigma E(\mathbf{q}_i^2) - \Sigma[E(\mathbf{q}_i)]^2 = E(\Sigma\mathbf{q}_i^2) - C[E(\mathbf{q})] = E[C(\mathbf{q})] - C[E(\mathbf{q})]$. Taking into account that $E[G(\mathbf{q})] = 1 - E[C(\mathbf{q})]$ and $G[E(\mathbf{q})] = 1 - C[E(\mathbf{q})]$ it follows that $\text{VAR}(\mathbf{q}) = G[(E(\mathbf{q})] - E[G(\mathbf{q})])$.

[13] The *pdf*'s $f(q_1)$ and $f'(q_1)$ correspond to two different cognitive contexts. A researcher X adopting $f(q_1)$ is inclined to believe that one of the objective probabilities $\mathbf{q}_1$, $\mathbf{q}_2$ is far higher than the other, though he is completely uncertain *which* one is higher. This uncertainty is revealed by the equality $E(\mathbf{q}_1) = E'(\mathbf{q}_2)$. For instance, consider the case where the objective probabilities that a tossed coin turns up heads or tails are represented by $\mathbf{q}_1$ and $\mathbf{q}_2$, respectively. If the coin appears noticeably irregular and X is totally uncertain whether the heads or the tails will profit from this irregularity, then he would presumably adopt a *cdf* such as $f(q_1)$. On the contrary, X will adopt a prior such as $f'(q_1)$ whenever he is inclined to believe that $\mathbf{q}_1$ and $\mathbf{q}_2$ have very close values.

[14] Clause (a) follows from the equalities $E[C(\mathbf{q})] = E(\Sigma\mathbf{q}_i^2) = \Sigma E(\mathbf{q}_i^2)$ and $E(\mathbf{q}_i^2) = p(Q_iQ_i)$ (cf. (33)).

Clause (b) follows from (a) taking into account that $E[G(\mathbf{q})] = 1 - E(C(\mathbf{q}))$.

[15] Analogously, the uniform distribution on the parameter vector **q** - which attributes equal density $f(q_1,...,q_{k-1}) = 1$ to all the values of **q** - determines a direct generalization of Laplace's rule, i.e., the inductive method characterized by the special values: $p(Q_i/e_n) = (n_i + 1)/(n + k)$. The evolution of the rule of succession from its original formulation (Laplace, 1774) to its recent generalizations is masterly traced by Zabell (1989).

[16] Clause (a) follows directly from (48).

Clause (b) can be proved as follows.

Given the non-negativity of the marginal variances var(q_i | e) derived from $F(q_1,...,q_{k-1})$ (cf. (20)(a)), it follows from the 'conditional formulation' of (35) that $p(Q_iQ_i/e) - [p(Q_i/e)]^2 \geq 0$. From this and the equality $p(Q_iQ_i/e) - [p(Q_i/e)]^2 = p(Q_i/e)[p(Q_i/eQ_i) - p(Q_i/e)]$ it follows that $p(Q_i/e)[p(Q_i/eQ_i) - p(Q_i/e)] \geq 0$. From this inequality, assuming that $p(Q_i/e)$ is positive, it follows that $p(Q_i/eQ_i) - p(Q_i/e) \geq 0$ and, equivalently, $p(Q_i/eQ_i) \geq p(Q_i/e_n)$ where the final inequality expresses the requirement of non-negative relevance (NNR). Lastly, when $p(Q_i/e) = 0$, then (NNR) is trivially satisfied. *Q.E.D.*

CHAPTER 4

[1] More generally, if a loss function such as $h(\theta - a)^2$ - where $h > 0$ - is used, the solution to the decision problem remains unaltered.

[2] See Niiniluoto (1986; 1987, Ch. 12.5) and Festa (1986).

[3] The use of the quadratic distance $D_2(a,\theta)$ appears to be appropriate in situation where the cognitive loss of accepting a, when the truth is θ, is not equal to the geometrical distance $|a - \theta|$ but increases in a way *more than proportional* to $|a - \theta|$.

[4] According to Lewis (*ibid.*, p. 62), the principle of immodesty is a necessary condition for a "*stable trust*" in a given inductive method: "If you wish to maximize accuracy in choosing an inductive method, and you have knowingly given your trust to any but an immodest method, how can you justify staying with the method you have chosen? If you really trust your method, and you really want to maximize accuracy, you should take your method's advice and maximize accuracy by switching to some other method that your method recommends. If that method also is not immodest, and you trust it, and you still want to maximize accuracy, you should switch again; and so on, unless you happen to hit upon an immodest method. *Immodesty is a condition of adequacy because it is a necessary condition for stable trust.*" [Author's italics].

[5] For more details on Lewis's approach to EPO for C-systems see Chapter 9.2.

[6] Note that the expected value $E[D_2(E(\mathbf{y}),\boldsymbol{\theta})]$ may be interpreted as the $p(\theta)$-based estimate of the distance between the future estimates of $\boldsymbol{\theta}$ made by the $p(\theta)$-based M-estimator E and the truth. Similarly, $E[D_2(T(\mathbf{y}),\boldsymbol{\theta})]$ may be interpreted as the $p(\theta)$-based estimate of the distance between the future estimates of $\boldsymbol{\theta}$ made by the estimator T and the truth.

Moreover, it should be pointed out that, *if* the immodesty of priors is defined w.r.t. the corresponding $p(\theta)$-based *M-estimates* $E(\mathbf{y})$ (see definition (7)), then the *quadratic distance* $D_2(E(\mathbf{y}),\boldsymbol{\theta})$ between such M-estimates and the truth may be justifiably used since such a distance function is the rationale for using M-estimators (see Section 1).

[7] Recall that the prior predictive probability $p(y)$ is derived from the available $p(\theta)$ and $p(y/\theta)$ applying formula (3.27).

[8] This is because all the properties used to prove (9) apply to any *cdf*. Similarly, definition (7) and theorems (9) and (10) can be extended to any conditional *cdf* $F(\theta/e)$ where e denotes an arbitrary background knowledge.

[9] Indeed Spielman proves that all C-systems are immodest. Moreover, Spielman (*ibid.*, p. 377) argues that "it can be shown by similar reasoning that all inductive methods that satisfy de Finetti's condition of exchangeability are immodest." More generally, Horwich (1982, p. 87) argues that "*every* inductive practice is immodest." However, Horwich's results apply only to "inductive practices" concerning *discrete* parameters.

[10] Indeed there is now an extensive and ever increasing literature on verisimilitude. To reach a rather exhaustive knowledge of this topic one might read the three books on the subject published so far: two treatises (Niiniluoto, 1987a; Oddie, 1987) and a collection of papers containing "a parade of approaches to truthlikeness" (Kuipers, 1987a). For a short survey of the current state of the debate concerning verisimilitude see Brink (1989) and Miller (1990).

[11] An exception is represented by Schurz and Weingartner (1987) who show that Popper's consequence approach can be rehabilitated by imposing appropriate restrictions on the classical notion of consequence-class of a theory used in Popper's original definition. More specifically, they (1) eliminate 'irrelevant' and 'redundant' consequences from the consequence-class so to obtain the so-called class of relevant consequence-elements and (2) apply to *this* class Popper's definition of closer-to-the-truth. They prove that this strategy blocks the otherwise fatal objections of Tichý and Miller against Popper.

[12] The above mentioned approaches to the explication of verisimilitude have been applied to many kinds of hypotheses considered in empirical sciences. For instance, the similarity approach initially was applied to hypotheses stated in propositional logic and in monadic first-order logic but later extensions covered full first-order logic, higher-order and intensional logics, probabilistic hypotheses, and hypotheses expressed in quantitative languages (for more information see Niiniluoto (1987, p. 198).

[13] The 'Bayesian estimation-approach' has also been applied to the analysis of appropriate procedures for point and interval estimation (see Niiniluoto 1986, 1987, Ch. 12.5; Festa, 1986).

[14] For a detailed exposition of the Bayesian approach to the epistemic problem of verisimilitude see Niiniluoto (1987, Ch. 7 and 12).

[15] In particular, see Kuipers (1992).

CHAPTER 5

[1] De Finetti (1937, pp. 124-129) proved his representation theorem for the case where $k = 2$. Subsequently, DFRT has been generalized to $k > 2$ (cf. Good, 1965, pp. 21-23).

[2] Indeed the equality used in (2)(a) to derive p_l from $F_l(q_1 \cdot \ldots \cdot q_{k-1})$ is the same as that used in (3.48) to derive a predictive distribution from a probability distribution on **q**.

[3] Cf. Hintikka (1971, pp. 329-336).

[4] Cf. Gaifman (1971, § 3) and Fine (1973, p. 194). See also Jeffrey's *Editor's Note* to Carnap (1980, § 20, p. 120).

[5] De Finetti conceived his representation theorem as a tool to get rid of unknown objective probabilities and, accordingly, considered probability distributions on unknown objective probabilities as "mere mathematical fictions" (Hintikka, 1971, p. 333).

[6] It may also be assumed that, in certain cases, *X* believes that the 'multivariate Bernoulli model' provides a convenient approximation to (idealization of) the real 'structure' of the multicategorical process *Ex* in examination.

[7] For the 'convergence towards the truth' of predictive probabilities see Hintikka (1971, pp. 336-339). See also Niiniluoto (1980, pp. 433-434 and p. 454, note 34).

CHAPTER 6

[1] The terms "Generalized Carnapian systems" and "GC-systems" - together with many other terms, symbols and definitions used in this section - are borrowed from Kuipers (1978, 1980).

[2] As Pietarinen (1972, pp. 60-61) points out, C-systems were anticipated in 1925 by W. E. Johnson in a paper published posthumously in 1932.

[3] The requirement (RR) also implies that the predictive probabilities do not depend upon the *variety* of evidence e_n. More precisely, (RR) implies that the special values $p(Q_i/e_n)$ do not depend upon the number *c* of different kinds of individuals exemplified in e_n, i.e., the number *c* of categories Q_i such that $n_i > 0$.

A universal hypothesis *h* = "All the individuals in the population *U* belong to one of the categories $Q_{i1},\ldots,Q_{ig}$" can be called a *genuine universal generalization* when $Q_{i1},\ldots,Q_{ig}$ is a *proper* subset of $\{Q_1,\ldots,Q_k\}$. It is well known that in an infinite population *U* the prior probability $p(h)$ and, therefore, the posterior probability $p(h/e_n)$ assigned to a genuine universal generalization *h* by C- and GC-systems are zero. As pointed out by Niiniluoto (1976, p. 428ff.), the construction of inductive methods which attribute non-zero prior probabilities to genuine universal generalizations requires the rejection of (RR) and the adoption of requirements of adequacy whereby the predictive probabilities $p(Q_i/e_n)$ depend not only on n_i and *n* but also on the variety of evidence *c*. Indeed requirements of this kind have been adopted by Hintikka and Niiniluoto (1974), Niiniluoto (1976), and Kuipers (1978, Ch. 6).

[4] At the turn of the century several students - including Hardy (1889), Withworth (1897, pp. 224-225), Gini (1911) and Lidstone (1920) - suggested that Beta distributions might be the appropriate priors for the Bayesian analysis of binomial inferences: see Good, 1965, p. 17;

Zabell, 1982, p. 1092; and Costantini, 1978, pp. 313-314; 1984, p. 154. For a description of the main features of Beta distributions see Wilks (1962, pp. 173-175).

5 The uniform distribution for $\mathbf{q}_1$ is represented by the *pdf* "$f(q_1) = 1$" which attributes equal density to all possible values of $\mathbf{q}_1$.

6 Lidstone (1920) was probably the first to suggest to use Dirichlet distribution in multinomial inferences (see Good, 1965, p. 25). For an exposition of the main features of Dirichlet distributions see Wilks (1962, pp. 177-182).

7 The uniform distribution on $\mathbf{q}$ is represented by the *pdf* "$f(q_1,\ldots,q_{k-1}) = 1$" which attributes equal density to all possible values of $\mathbf{q}$.

8 For a long time the equivalence between GC-systems on the one hand and Beta and Dirichlet distributions on the other, apparently was not noticed. As far as I know this equivalence was pointed out first by Good (1965, p. 17): "It seems possible that G. F. Hardy [who defined Beta distributions] was the first to suggest a 'continuum of inductive methods', to use Carnap's phrase." Later, the equivalence between GC-systems and Dirichlet (Beta) distributions was proved by Jamison (1970, pp. 47-49), Rosenkrantz (1977, pp. 71-73), and others.

9 The only difference is that $p_0(Q_i/e_n) = n_i/n$ is not defined in the case where $n = 0$, whereas it follows from (34) that $p_{\gamma,0}(Q_i/e_0) = p_{\gamma,0}(Q_i) = \gamma_i$.

10 Several arguments against the straight rule and $(\gamma,0)$ have been given (see Carnap, 1952, § 14; 1980, pp. 85-86 and 95; Good 1965, pp. 15-18, 23, 28 and 36; Kuipers, 1978, p. 52). Note, for instance, that the apparently plausible requirement of adequacy (Reg) (see Chapter 2.1) is not satisfied by extreme inductive methods $(\gamma,0)$ since they attribute prior probability zero to any 'heterogeneous' sequence e_n, i.e., to any sequence containing more than one kind of trial.

11 In (38) the term "GC-systems" is interpreted to include the extreme GC-systems $(\gamma,0)$ and (γ,∞) with λ-values $\lambda = 0$ and $\lambda = \infty$. The class of 'proper' GC-systems, where $0 < \lambda < \infty$, is obtained by adding (i) the axiom of regularity (Reg), which leads to the exclusion of $\lambda = 0$, and (ii) the axiom of positive relevance (PR), which leads to the exclusion of $\lambda = \infty$.

12 Johnson (1932, pp. 421-423) anticipated not only Carnap's C-systems (cf. note 2) but also their axiomatizations: see Good (1965, pp. 25-26) and Zabell (1982).

13 Indeed (RL) - as opposed to (RR), which becomes tautological when $k = 2$ - imposes a genuine restriction on the predictive probabilities $p(Q_i/e_n)$ also when $k = 2$ (cf. Good, 1965, p. 26).

14 It follows from (39) that a unified axiomatization of GC-systems would consist in adopting the requirement of linearity (RL) for *any* $k \geq 2$. However, a serious defect of this axiomatization is that the intuitive meaning of (RL) is much less clear than that of the requirement of restricted relevance (RR) used in (38) (cf. Carnap, 1980, pp. 101-102).

15 Note that (CIRQ), if applied when $k > 2$, immediately excludes any analogy by similarity. However (CIRQ), as opposed to (RR), enacts a genuine restriction on the predictive probabilities $p(Q_i/e_n)$ also when $k = 2$.

16 A possible criticism of Costantini's axiomatization is that the new axiom (CIRQ) is less intuitively plausible than the 'old' axiom (RR).

17 Besides the objective probabilities governing a multivariate Bernoulli process, many other types of proportions are considered in the empirical sciences. In fact, proportions are of interest whenever "some unit is broken into parts, such as proportions (by weight) of various chemical constituents of a substance" (Connor and Mosimann, 1969, p. 194). For some examples concerning the biological sciences see Mosimann (1962, pp. 77-81) and Connor and Mosimann (1969, pp. 200-205).

Given the constraint $\Sigma \mathbf{q}_i = 1$, the proportions $\mathbf{q}_1,...,\mathbf{q}_k$ cannot be independent of each other in the usual sense (as defined in (3.4)); indeed "statisticians and biologists alike have been wary of the correlations existing among proportions or percentages since Pearson's (1897) paper on spurious correlations" (Mosimann, 1962, p. 65). However, very often these correlations do not denote any 'genuine' correlation among the proportions but simply "serve as measures of the correlation due to the constraint [$\Sigma \mathbf{q}_i = 1$]" (Connor and Mosimann, 1968, p. 78).

The concept of neutrality aims to individuate those cases where there is no genuine correlation among the proportions $\mathbf{q}_1,...,\mathbf{q}_k$. More precisely: when all the proportions $\mathbf{q}_i$ are *neutral* w.r.t. $\mathbf{q}=(\mathbf{q}_1,...,\mathbf{q}_k)$, then the correlations existing between the different couples of proportions depend *uniquely* on the constraint $\Sigma \mathbf{q}_i = 1$.

18 Other 'neutrality-based' parametric requirements conceptually akin to (CM) have been used to formulate other axiomatizations of Dirichlet distributions (Darroch and Ratcliff, 1971; Doksum, 1974, pp. 187-188; and James and Mosimann, 1980).

CHAPTER 7

1 For a survey of interviewing procedures see Winkler (1972, pp. 182-192).

2 The mean of $\mathbf{q}_1$ can be seen as X's estimate of $\mathbf{q}_1$ and the variance of $\mathbf{q}_1$ as X's uncertainty about his own estimate (cf. Chapter 4.1). Since "it is very difficult to assess the variance of a distribution intuitively" (Winkler, *ibid.*, p. 189), more feasible methods of fitting a Beta distribution should be considered (see Winkler, *ibid.*, p. 188-189 and Novick and Jackson, 1974, pp. 167-170).

3 A well known example of such paradoxes is the so-called Bertrand's Paradox (see Rosenkrantz, 1977, pp. 73-74).

4 A concise exposition of Jeffreys's theory of invariance and related results is given by Zellner (1971, pp. 41-53). See also Barnett (1974, p. 178-179) and Seidenfeld (1979a, p. 417-423).

[5] In one sense the problem of 'arbitrariness of the group of transformations' is analogous to the problem of arbitrariness of parametrization: see Fine (1973, p. 172) and Milne (1983, p. 52).

[6] Indeed, in principle it cannot be excluded that certain invariance requirements are also satisfied by intuitively 'very informative' priors. This would imply that invariance properties are not sufficient to identify appropriate ignorance priors.

[7] See, for instance, Perks (1947), Hartigan (1964), Villegas (1977), Novick and Hall (1965) and Novick (1977).

[8] An excellent account of Jeffreys's theory of Invariants is given by Huzurbazar (1976, part 1).

[9] Box and Tiao (1973, p. 25) argue that "a prior distribution is supposed to represent knowledge about parameters before the outcome of a projected experiment is known. Thus, the main issue is how to select a prior which provides little information relative to what is expected to be provided by the intended experiment." An even more radical formulation of the view that prior distributions should be experiment-dependent is given by Skilling (1985, p. 13) who points out that "different experiments can demand different priors for their interpretation. It ... follows that the prior *cannot* be identified with the current state of knowledge, as we have assumed until now." Rather the prior should be "identified with the *question being asked* of the experiment" (*ibid.*, p. 15).

[10] Bernardo (1979) offers an interesting analysis of reference priors and also gives an extensive bibliography on the subject.

[11] Usually a significant proportion of the background knowledge *BK* is 'tacit knowledge' which can be made explicit by self-interrogation and critical discussion.

[12] For an introductory account of the 'meaning' and the properties of cross entropy see Rosenkrantz (1981, Ch. 4.3 and 4*.3). As pointed out by Williams (1980, p. 133, note 1), the quantity $I_K(f_2, f_1)$ and other expressions of this type have already been studied by Good (1950), Kullback and Leibler (1951), Savage (1954) and Lindley (1956).

[13] A formulation of this principle has already been given by Good (1963): cf. Williams (1980, p. 134, note 1). On the origins of MAXENT and, more generally, the so-called information-theoretic statistics see, for example, Guiasu (1977, Ch. 17), Jaynes (1979, pp. 16-44) and Rosenkrantz's introduction to Jaynes (1983).

[14] For the interpretation of distributional constraints, see Seidenfeld (1979a, pp. 423 and 438, n. 20).

[15] For instance, Shimony (1985), Seidenfeld (1979a, 1986) and other authors claim that (PME) may leads to the selection of unacceptable prior distributions.

[16] Claim (1) can be seen as a particular case of a more general 'least change principle' proposed for dealing with the so-called probability kinematics (cf. May and Harper, 1976, p. 139-140 and Domotor, 1985, pp. 75-76).

[17] For example, alternative measures of distance between (epistemic) probability distributions are considered by May and Harper (1976, p. 145), Domotor (1985, pp. 88-90) and Gaifman (1986, pp. 325-331).

[18] The larger X's inductive inertia λ the slower X modifies his a priori vector γ in response to empirical evidence e_n (cf. (6.4)).

[19] Some ambiguities of Carnap's views concerning the nature and role of inductive intuition are pointed out by Giere (1975, p. 195): "The axioms are justified to the extent that the derived consequences agree with 'our' (Whose? How determined?) intuitions concerning which actions, beliefs, etc. are rational in these possible cases. ... The appeal to such intuitive principles raises several important philosophical questions. For example, What gives such principles normative rather than merely descriptive force? ... Can one not ... ask whether the principles 'we' in fact accept are *really* correct principles of rational behaviour?"

[20] Carnap's argument includes the following weak point. Although (RR) implies η-equality, (RR) *is not* implied by η-equality. Hence, Carnap's assumption that (RR) should be adopted whenever η-equality is assumed is not compelling.

Consider, for instance, the class of inductive methods introduced by Hintikka and Niiniluoto (1976). While they do not satisfy (RR) it would appear that they satisfy η-equality. For a concise description of such methods see Kuipers (1978, pp. 71-75; 1980, pp.188-189).

[21] It is clear from the context that Carnap's preference for $\lambda = 1$ also applies to equal-distance families.

[22] Let the properties Yellow and Orange be denoted by Y and O. Then, the 'negative (inductive) analogy' between Y and O could be expressed by the inequality: $p(O/e_n Y) < p(O/e_n)$, where "O" denotes the hypothesis that the next observed car will be Orange, and "$e_n Y$" denotes the evidence obtained by adding the observation of a yellow car to the previously observed sample e_n randomly drawn from the population of cars. Analogously, the 'positive (inductive) analogy' between Yellow and Orange could be expressed by the inequality $p(O/e_n Y) > p(O/e_n)$.

[23] I maintain that in the example described herein, requirement (RR) should not be accepted given that *BK* reveals the existence of a 'negative analogy' between Yellow and Orange. Note that the reasons for rejecting (RR) are the opposite to those which would be suggested by Carnap who would have referred to the (alleged) universally valid 'positive analogy' between Yellow and Orange.

[24] It should also be pointed out that if the value $\lambda = 1$ - tentatively proposed by Carnap as the optimum λ-value - is adopted then $E[G(\mathbf{q})] = [\lambda/(\lambda + 1)]G(\gamma) = (1/2)G(\gamma)$. In other words, if $\lambda = 1$ is adopted, then the expected disorder of the population under investigation is equal

to half the maximum value $G(\gamma)$. Once again it is difficult to see why this strong a priori belief should be adopted in *any* empirical research.

[25] Note that this distribution, being the only one to satisfy the constraints included in I, is also - very trivially - the MAXENT distribution. This suggests that a priori inductive principles such as (PME) or (MCB) are dispensable whenever appropriate distributional and formal constraints are adopted. A different example where a formal constraint is sufficient to determine the uniquely admissible probability distribution is considered by Spielman (1981, p. 358).

[26] In clause (i) it is assumed that the background knowledge *BK* shared by a given group of researchers also includes some more or less explicitly stated 'standards of similarity' to be used in the identification of the experimental processes 'sufficiently similar' to *Ex*.

[27] The term "external" is used because $Ein(\mathbf{\theta},Ex)$ does not describe any experimental evidence obtained from any experiment **y** related to $\mathbf{\theta}$ (in *Ex*) by a likelihood function $p(y/\theta)$.

[28] Note that the two formal properties stated in (Ext) are also satisfied by expected values and several other 'formal' estimates used in statistics.

[29] This involves, among other things, the supposition that the ratios $a_j^t/(1 - a_i^t)$ *do not* display any appreciable tendency to increase - or decrease - when the corresponding values of a_i^t increase.

[30] Although a formal definition of "$H_1,...,H_m$-independence of Q_j from Q_i" could be given, it is not indispensable for our purposes, since a careful inspection of the two sets $a_i^1,...,a_i^m$ and $a_j^1/(1 - a_i^1),...,a_j^m/(1 - a_i^m)$ appears sufficient to recognize the presence of an appreciable dependence between them.

[31] Recall that $\mathbf{q}_j/(1 - \mathbf{q}_i)$ is independent of $\mathbf{q}_i$ when the probabilities attributed to the possible values of $\mathbf{q}_j/(1 - \mathbf{q}_i)$ are not affected by information about the value of $\mathbf{q}_i$ (see Chapter 6.6).

[32] Given that all GC-systems satisfy the requirement of restricted relevance (RR), my contextual justification of (CM) (indirectly) provides a contextual justification of (RR).

CHAPTER 8

[1] The restriction "if m is not too low" on the estimation procedures (1) and (2) for $Est(\mathbf{q}_i)$ can be justified as follows.

When the only external information $EIn(\mathbf{q},Ex)$ in *BK* is given by the experimental information $H_1,...,H_m$, where m is *very low* - let us say $m = 1, 2,$ or 3 - then there are no grounds for believing that the specific values of $\mathbf{q}$ specified by $H_1,...,H_m$ are the result of some 'causal mechanism' producing $Ex_1, Ex_2, ..., Ex_m$ and the other *Ex*-similar processes and not just a coincidence. Hence, any external estimate $Est(\mathbf{q}_i)$ derived from $H_1,...,H_m$ - by the application of procedures such as (1) or (2) - is doomed to be extremely unreliable. (Cf. the analogous considerations made in Chapter 7.6, pp. 100-101, where it was claimed that the $H_1,...,H_m$-independence of Q_j and Q_i suggests the acceptance of "the independence of Q_i and Q_j" under the restriction that m is *sufficiently high*.)

In cases such as those above - and, more generally, in any case where the external information $EIn(\mathbf{q},Ex)$ included in BK is very poor - it would seem reasonable to select the symmetrical prior vector $\mathbf{1/k}=(1/k,...,1/k)$ as the external estimate $Est(\mathbf{q})=(Est(\mathbf{q}_1),...,Est(\mathbf{q}_k))$ of $\mathbf{q}$. Presumably, the same decision should be adopted in the cases of maximum ignorance, where BK does not contain any external information $EIn(\mathbf{q},Ex)$

[2] Already this decision can be seen as an application of the requirement of cognitive coherence (CC) which - given the external estimates $Est(\mathbf{q}_1),..., Est(\mathbf{q}_k)$ - imposes adoption of the following m-constraints $c_1,...,c_k$ on Dir(γ,λ): c_1 = "$E(\mathbf{q}_1) = Est(\mathbf{q}_1)$", ... , c_k = "$E(\mathbf{q}_k) = Est(\mathbf{q}_k)$".

[3] The restriction "if m is not too low" on the estimation procedures (3) and (4) for $Est[G(\mathbf{q})]$ is identical to the restriction suggested w.r.t. the procedures (1) and (2) for $Est(\mathbf{q}_i)$ and can be justified along the same lines (cf. note 2). Indeed, this restriction can be seen as the particular case of a more general restriction applicable to any procedure of the same type as (1)-(4), to be used for making an external estimate $Est(\boldsymbol{\theta})$ of the parameter $\boldsymbol{\theta}$ of a given process Ex.

The above mentioned restriction on (1)-(4) could be replaced by a more precise restriction such as: "The estimation procedures (1)-(4) can be applied only if $m \geq t$" where t denotes a certain preselected threshold value.

[4] The remarks on the procedures (1)-(4) (see notes 2 and 3) apply also to the restriction "if m is not too low" on the procedures (5) and (6) for $Est[D_2(\gamma°,\mathbf{q})]$.

Moreover, in the particular case where $m = 1$ - i.e., in the case where the available external information $EIn(\mathbf{q},Ex)$ in BK is given by the proposition H_1 which specifies the value $a^1=(a^1_1,...,a^1_k)$ of $\mathbf{q}$ in a certain Ex-similar process Ex_1 there is another reason for not using the estimation procedures (1) and (5).

Indeed, in the case where $m = 1$, one can see that: (i) the estimation procedure (1) leads to the external estimates $Est(\mathbf{q}_i) = a^1_i$ for any $\mathbf{q}_i$ and, consequently, to the selection of the optimum prior $\gamma° = Est(\mathbf{q}) = a^1$; (ii) the estimation procedure (5) implies that $Est[D_2(\gamma°,\mathbf{q})] = D_2(\gamma°,a^1)$. From the equalities $\gamma° = a^1$ and $Est[D_2(\gamma°,\mathbf{q})] = D_2(\gamma°,a^1)$ it follows that $Est[D_2(\gamma°,\mathbf{q})] = D_2(a^1,a^1) = 0$. The external estimate $Est[D_2(\gamma°,\mathbf{q})] = 0$ can be seen as a formal representation of the guess that the external estimate $Est(\mathbf{q})$ (= $\gamma°$) is exactly true. However, it is pretty clear that the above guess is unjustifiably optimistic, since $Est(\mathbf{q})=(Est(\mathbf{q}_1),...,Est(\mathbf{q}_k))$ is based on the extremely poor experimental information H_1. (Analogously, one can prove that, in the case where $m = 1$, the procedures (2) and (6) also lead to the unreliable external estimate $Est[D_2(\gamma°,\mathbf{q})] = 0$.)

One may wonder how an external estimate $Est[D_2(\gamma°,\mathbf{q})]$ could be made in the cases where m is very low and, more generally, in the cases where the external information $EIn(\mathbf{q},Ex)$ included in BK is very poor. An apparently plausible possibility is suggested by the circumstance that $Est[D_2(\gamma°,\mathbf{q})] = Est[D_2(Est(\mathbf{q}),\mathbf{q})]$ may be interpreted as an external estimate of the 'inaccuracy' (distance from the truth) of the external estimate $Est(\mathbf{q}) = \gamma°$. Indeed, this interpretation appears to suggest that the value of $Est[D_2(\gamma°,\mathbf{q})]$ should increase when the 'size' (and the 'reliability') of the external information $EIn(\mathbf{q},Ex)$ - from which $Est(\mathbf{q})$ is derived - decreases. In particular, $Est[D_2(\gamma°,\mathbf{q})]$ should reach its maximum value in those situations where the 'size' of $EIn(\mathbf{q},Ex)$ is minimum (see, for instance, the case of maximum ignorance described in the last paragraph of note 2).

The above remark leads to the conclusion that any multinomial context where an external estimate $Est(\mathbf{q})=Est(\mathbf{q}_1),...,Est(\mathbf{q}_k)$ of $\mathbf{q}$ can be made - i.e., virtually any multinomial context (cf. note 2) - is a D-context.

5 Since all the possible values of the quadratic distance $D_2(\gamma°,\mathbf{q})$ (= $\Sigma(\gamma_i^\circ - \mathbf{q}_i)^2$) are non-negative, it follows from (Ext)(b) that $Est[D_2(\gamma°,\mathbf{q})] \geq 0$, i.e., that the minimum value of $Est[D_2(\gamma°,\mathbf{q})]$ is 0. Analogously, since all the possible values of $G(\mathbf{q})$ (= $1 - \Sigma\mathbf{q}_i^2 = \Sigma q_i(1 - q_i)$: cf. the paragraph under formula (10.12), p. 142) are non-negative, it follows from (Ext)(b) that $Est[G(\mathbf{q})] \geq 0$, i.e., that the minimum value of $Est[G(\mathbf{q})]$ is 0. Now, from the fact that the minimum value of $Est[D_2(\gamma°,\mathbf{q})]$ and $Est[G(\mathbf{q})]$ is 0, it follows - taking into account that, due to (7), $Est[D_2(\gamma°,\mathbf{q})] + Est[G(\mathbf{q})] = G(\gamma°)$ - that the maximum value of both the external estimates $Est[D_2(\gamma°,\mathbf{q})]$ and $Est[G(\mathbf{q})]$ is $G(\gamma°)$.

6 It can be proved, indeed, that the equality in (7) is implied by (1), (3) and (5). To this purpose consider the following chain of equalities:

$$Est[D_2(\gamma°,\mathbf{q})] = \frac{\sum_{t=1}^{m}\sum_{i=1}^{k}(\gamma_i^\circ - a_i^t)^2}{m} \text{ (from (5))} = \frac{\sum_{i=1}^{k}\sum_{t=1}^{m}(\gamma_i^\circ - a_i^t)^2}{m}$$

$$= \frac{\sum_{i=1}^{k}[m(\gamma_i^\circ)^2 + \sum_{t=1}^{m}(a_i^t)^2 - 2\gamma_i^\circ\sum_{t=1}^{m}a_i^t]}{m} = \frac{\sum_{i=1}^{k}[m(\gamma_i^\circ)^2 + \sum_{t=1}^{m}(a_i^t)^2 - 2m(\gamma_i^\circ)^2]}{m}$$

(since it follows from (1) that $\sum_{t=1}^{m} a_i^t = mEst(\mathbf{q}_i) = m\gamma_i^\circ$)

$$= \frac{\sum_{t=1}^{m}\sum_{i=1}^{k}(a_i^t)^2 - m\sum_{i=1}^{k}(\gamma_i^\circ)^2}{m} = \frac{\sum_{t=1}^{m}[C(a^t)] - mC(\gamma°)}{m}$$

$$= \frac{\sum_{t=1}^{m}[1 - G(a^t)] - m[1 - G(\gamma°)]}{m} = \frac{mG(\gamma°) - mEst[G(\mathbf{q})]}{m} \text{ (from (3))}$$

$$= G(\gamma°) - Est[G(\mathbf{q})], \ Q.E.D.$$

7 In the proof, which is similar to that given in the note above, it is assumed that the same weights $w_1,...,w_m$ are used in (2), (4) and (6). This assumption is plausible since in all three

cases the same experimental information $H_1,...,H_m$ is used and the weights $w_1,...,w_m$ are related to certain features of $H_1,...,H_m$ (cf. the remark below (2)).

[8] In the last paragraph of note 4 it was claimed that virtually any multinomial context is a D-context. Since any D-context is also a GD-context, this claim implies that virtually any multinomial context is a GD-context.

[9] However, a statistician working on Bayesian elicitation might remark that a GD-context provides 'better' (i.e., more stable) results, because the researcher who finds that (7) is violated may wish to reconsider his external estimates - at least for $D_2(\gamma°,\mathbf{q})]$ and $G(\mathbf{q})]$ and possibly also for $\mathbf{q}_I$ - in order to come to a compromise fulfilling (3). This strategy is in line with the results obtained by many researchers (such as Novick and Jackson (1974), Hogarth (1980), Lourens (1984), Terlouw (1989)) who found that some overdetermination of the prior is very useful: inconsistencies lead to reconsideration, and consistencies give more trust in the results. This remark was suggested by Prof. Molenaar.

[10] Note, in particular, that the CC-optimum value λ_{CC} becomes equal to 0 when $Est[D_2(\gamma°,\mathbf{q})]$ reaches its maximum value $G(\gamma°)$ (cf. note 5) while it tends to ∞ when $Est[D_2(\gamma°,\mathbf{q})]$ tends to its minimum value 0. This suggests the integration of the CC-solution by putting $\lambda_{CC} = \infty$ in the case where $Est[D_2(\gamma°,\mathbf{q})] = 0$.

[11] Note that the solution for high values of $Est[D_2(\gamma°,\mathbf{q})]$ implied by (9) can be seen as a particular case of the choice of diffuse priors, usually suggested in Bayesian statistics, for those situations where there is very scarce background knowledge about the value of the parameter under investigation.

[12] Notice that $D(T(\boldsymbol{\theta} \mid y),\theta)$ depends both on the value θ of $\boldsymbol{\theta}$ and on $T(\boldsymbol{\theta} \mid y)$ which, in turn, depends both on the estimator T and the result y of the experiment **y** which has been performed. Whilst **y** and T are selected by X, the *actual* state of nature θ and the *actual* experimental result y are, as it were, selected by 'the world'.

[13] See the justification for using the quadratic distance as provided in Chapter 4, note 6.

[14] Theorem (22) can be proved as follows.
$E(D_2(\nu(\mathbf{e}_n),q) \mid q) = E[\Sigma\ (n_i/n - q_i)^2] = \Sigma[E(n_i/n - q_i)^2)] = \Sigma[\text{var}(n_i/n) + [E(n_i/n - q_i)]^2]$ (from (3.20.b) and (3.20.c)). From these equalities, taking into account that $E(n_i/n) = q_i$ (from (3.24) and (3.16.a)) and, consequently, $[E(n_i/n - q_i)] = 0 = [E(n_i/n - q_i)]^2$ (from (3.16.a)), it follows that $E(D_2(\nu(\mathbf{e}_n),q) \mid q) = \Sigma\text{var}(n_i/n) = \Sigma[nq_i(1 - q_i)/n^2]$ (from (3.25)) $= \Sigma[q_i(1 - q_i)/n]$ $= (1 - \Sigma q_i^2)/n = G(q)/n$.

[15] It should be recalled that, analogously to all formulae in Chapter 3.2, formula (3.19) does not apply only to parameters and parameter vectors but also to random variables and mixtures of parameter vectors and random variables (cf. p. 23). Hence, it also applies to any probability distribution on $(\mathbf{q},\mathbf{e}_n)$ where $\mathbf{q}$ is a parameter vector and $\mathbf{e}_n$ a random variable with a probability distribution which depends on the value of $\mathbf{q}$.

[16] According to Carnap (*ibid.*, pp. 70-71) the quantity $\Sigma q_i^2 = C(q)$ occurring in (30) can be seen as a measure of "the degree of order or uniformity or regularity" of a population. In this connection, he points out that (i) in the case of a *completely homogeneous* population - i.e., when one q_i is 1 and all others are 0 - the value of Σq_i^2 is maximum ($\Sigma q_i^2 = 1$) while (ii) in the case of a *maximally heterogeneous* population - i.e., when $q_i = 1/k$ for all q_i - the value of Σq_i^2 is minimum ($\Sigma q_i^2 = 1/k$).

Apparently Carnap stated his interpretation of Σq_i^2 without being aware that Σq_i^2 had already been proposed as a plausible uniformity index by several statisticians and scientists (cf. Chapter 10).

[17] A number of modifications to Carnap's terminology are made to simplify the analysis of such relationships.

[18] Carnap's decision to use the square error as a measure of the seriousness of the error was in keeping with a long statistical tradition which, in turn, was motivated to a large extent by the mathematical tractability of the square error. However, Carnap did not provide any *theoretical* reason for his proposal. For a possible 'justification' of the square error, or quadratic distance from the truth, see Chapter 4, note 6.

[19] Carnap and Stegmüller introduced GC-systems only in (1959).

[20] Concerning these differences between my approach and that of Carnap it should be recalled that in 1952 the equivalence between symmetrical Dirichlet distributions and C-systems was still to be recognized and the theory of verisimilitude was still to be developed.

CHAPTER 9

[1] Here Carnap would appear to imply that λ-based inductive estimates of the success of inductive methods cannot be used to select the optimum inductive method. Carnap may have supposed, rightly, that any inductive method would immodestly evaluate itself as optimum.

[2] This mistake was identified by Spielman (1972, p. 376) and immediately accepted by Lewis (1974, p. 84).

[3] This is somewhat problematical: if the prior depends on the data, then the posterior is no longer proportional to the *product* of the prior and the likelihood: therefore Bayes's theorem no longer holds. This observation was made by Prof. Molenaar.

[4] This particular aspect of my approach to EPO appears to meet a desideratum expressed by Pietarinen (1974). Indeed, concerning Lewis's hyperempiricist view, Pietarinen (1974, p. 197) remarks: "When the relative frequency of a kind [Q_i] in a sample is [n_i/n], the estimation formula [suggested by Lewis] recommends a certain method [λ]. But whether or not the method is rational depends on the situation, in particular, on the question of [the degree of uniformity of the universe]. This cannot be decided on the basis of the frequencies in the sample alone; it is tempting to say that somehow it must be *presupposed*. ... In Lewis's

procedure ... the only relevant evidential information consists of the size and the degree of uniformity of the sample. But ... to know these things is not enough to estimate the degree of uniformity of the universe. Ordinary inductive situations ... may differ in respect of [the degree of uniformity], and this is relevant for the choice of the value of λ." According to Pietarinen (1974, p. 198) both Carnap's aprioristic view and Lewis's hyperempiricist view "are defective" since they are unable to take account of "the degree of uniformity of the things under study". On the contrary, my contextual approach to EPO shows a possible way of considering the estimated degree of (dis)order of the universe under examination.

5 This procedure had already been described by Kuipers (1984) who analyzed the limit-process in the special case where $\lambda_0 = 0$.

6 Remember that the 'extreme' λ-values are $\lambda = 0$ and $\lambda = \infty$ (see Chapter 6.4). Kuipers (*ibid.*, pp. 40-41) also shows that, for some possible e_n and λ_0, the result of the limit-process is an extreme λ-value. Kuipers leaves unanswered the question whether or not this is a reason not to use certain values of λ_0 as possible starting values of the process.

7 There is a strong resemblance between my immodesty-based interpretation of Kuipers's limit-process and Lewis's description (1971, p. 62) of the behaviour of a person using a modest inductive method (cf. Chapter 4.2).

8 Remember that, for any $\lambda \neq \infty$, $p_\lambda(Q_iQ_i/e_n) > [p_\lambda(Q_i/e_n)]^2$: cf. Chapter 3, note 16. It would be interesting to ascertain whether some limit-process of the kind described by Kuipers could also be generated by the λ_0-based Bayesian estimates $E_{\lambda 0}(C(\mathbf{q}) \mid e_n)$. However, even if the non-Bayesian estimate in (1) were replaced by $E_{\lambda 0}(C(\mathbf{q}) \mid e_n)$, the whole procedure (1)-(2) would remain non-Bayesian. In fact, the usual Bayesian procedures allow calculation of the expected error - w.r.t. an inductive method λ_0 - of all inductive methods (including λ_0) relative to any evidence e_n. Hence, a Bayesian who 'starts' with method λ_0 and evidence e_n would select that method λ_1 which minimizes the expected error - relative to λ_0 and e_n - as the most appropriate inductive method. But theorem (4.11) implies that *any* inductive method λ_0 immodestly indicates itself as being maximally appropriate. This means, in Kuipers's terminology, that the 'estimate λ_1' suggested by the starting method λ_0 is ... λ_0!

9 Kuipers's second solution to EPO is *formally* similar to my V_1-solution (see Chapter 8.5). However, there is a substantial difference between the two solutions. Whilst $est(C(\mathbf{q}) \mid e_n)$ is a 'formal' e_n-based estimate, my external estimate $Est[C(\mathbf{q})]$ is an 'informal' estimate based on the background information *BK*.

10 A hyperempiricist view also appears to underlie Hintikka's assumption (1987, p. 305) that a researcher "might initially choose a large Carnapian index of caution λ. But if the overwhelming majority of observed individuals belongs to a small number of cells, [he] might very well be led to acknowledge that his caution was excessive and that he or she ought to have opted for a smaller λ in the first place."

Several proposals in line with the hyperempiricist view are advanced by Good (1965, pp. 28 and 35-38). They involve estimating the parameter α^* of a symmetrical Dirichlet distribution Dir(α^*,...,α^*) on the basis of the empirical sample e_n. For instance, if I understand Good (ib., p. 35) correctly, the value α^* might be selected so as to maximize the prior predictive probability $p(e_n)$ derived using formula (3.48) from Dir(α^*,...,α^*).

[11] Regarding the nature of the new probability system, Kuipers (*ibid.*) points out that "it is equally easy to see that the resulting system, for $k \geq 3$, cannot have the property that $p(Q_i/e_n)$ is only a function of n and n_i, being a structural property of all λ-systems". It is not clear to me whether such a system is exchangeable although I guess it is not. Note that in Kuipers's new system the special values $p(Q_i/e_n)$ depend, besides on n and n_i, only on the degree of order $\Sigma(n_i/n)^2$ of e_n.

[12] Rosenkrantz (1982, p. 93), referring to the family **Q** = (Q,non-Q), points out that even the extreme inductive methods $\lambda = 0$ and $\lambda = \infty$ can be adopted when certain 'presuppositions' about the degree of homogeneity of the considered population are made. Indeed he points out that "[$\lambda = 0$] does seem appropriate for somebody who knows that the population is completely homogeneous, and ... [$\lambda = \infty$] seems apt for one who *knows* that half the population is Q and half non-Q.". (Cf. the proposal made in Chapter 8, note 10.)

[13] For this purpose one could use some technical results obtained by Walk (1963, pp. 523-528 and 1966, pp. 76-78) who showed that "the optimal value of λ is a monotonic function of the entropy in one's universe" (Hintikka, 1987, p. 303).

[14] This view appears to be shared by Horwich (1982, p. 81) who claims that "a reasonable inductive practice must be *demonstrably reliable*: there must be reason to think that it is good at directing us towards the truth. ... So it can be reasonable to adopt one practice rather than another only if it can be shown that the one is more likely to be successful".

[15] Note that according to Friedman (*ibid.*, p. 369), "the kind of probability at issue is objective or physical probability. We are requiring that there be a lawlike statistical correlation between the property of being reached by [a given inductive method] and the property of being true, in the very same sense in which there is a lawlike statistical correlation ... between smoking and lung cancer." A strong argument against Friedman's formulation of the verisimilitude approach to scientific method is given by Niiniluoto (1980, pp. 453-454).

[16] Here "representative" is not used in the standard statistical sense regarding the sampling procedure used for drawing the sample from the population. Rather the sample is "representative of the entire universe" if its composition is similar to that of the universe.

[17] Rosenkrantz (1982, p. 93) points out that Carnap's preference for small values of λ can be justified only "by confining it to contexts where prior information about the population is lacking. It then becomes a special case of the principle favoring adoption of a diffuse prior in the absence of prior information."

[18] Rosenkrantz (1982, p. 93), indeed, points out that "the problem of induction ... is to justify that preference."

[19] Regarding the cases where the available background knowledge *BK* is very poor, see Chapter 8, note 4.

CHAPTER 10

[1] Already, between the end of the 1880's and the beginning of the 1900's, a number of scientists and statisticians - such as Quetelet, Galton, Karl Pearson, and Yule - were concerned with the *problem* of qualitative variability.

[2] David (1968, p. 573; 1981, p. 436) provides some details on the history of mean difference and related indices. For instance, he points out that mean difference was already being used as a measure of quantitative variability by German astronomers even as early as 1869.

[3] For instance, Good (1982, p. 562) - although being aware that "both the quadratic and the logarithmic index of heterogeneity [i.e., Gini diversity $1 - \Sigma q_i^2$ and Shannon entropy $H(q) = \Sigma q_i \log q_i$] have histories dating back a hundred years" - appears to ignore the importance of Gini's work and mentions only that "recently Bhargava and Uppuluri (1975) called $[1 - \Sigma q_i^2]$ the Gini index and cited Gini (1912)".

[4] Good (1982), having provided much information on the history of the homogeneity index Σq_i^2, maintains that "it is unjust to associate $[\Sigma q_i^2]$ with any one person" since "if $[q_1,...,q_k]$ are the probabilities of $[k]$ mutually exclusive and exhaustive events, any statistician of this century who wanted a measure of homogeneity would have taken about two seconds to suggest $[\Sigma q_i^2]$" (*ibid.*, p. 562). However, some interesting historical circumstances told by Good himself appear to be in conflict with his idea that the discovery of Σq_i^2 as a homogeneity index is totally trivial. Indeed, he states that the famous mathematician A. M. Turing - when, as head of the section of the Government Code and Cipher School in Bletchley, Buckinghamshire was working on the cryptanalysis of the German cryptographic machine Enigma - being "probably unaware of the previous uses [of Σq_i^2], gave $[\Sigma q_i^2]$ its most natural name, 'the repeat rate' ... and made very important use of $[\Sigma q_i^2]$ in cryptanalysis" (*ibid.*). Moreover, Good (*ibid.*) informs us that the biologist E. H. Simpson also worked at Bletchley: for this reason, when "he later suggested the use of 1 minus the repeat rate as an index of diversity he did not acknowledge Turing, fearing that to acknowledge him would be regarded as a breach of security".

[5] The term "dichotomous" is used "because the rarity of the species $[Q_i]$ would be unchanged if the other species were grouped into a single complementary category" (Patil and Taillie, 1982, p. 548).

[6] Cf. Patil and Taillie (1982, p. 551).

[7] For instance, Hill (1973, p. 428) remarks that Σq_i^2 represents a "weighted mean of the proportional abundances" q_i of the different species Q_i in a given population. Note that Weaver (1948), dealing with a process characterized by k possible events, already took Σq_i^2 as a measure of the "expected commonness" of an event (see Pielou, 1975, p. 95).

[8] The family $N_a(q)$ is strictly connected to the so-called *generalized entropy of order a*, or $H_a(q)$ - defined as $H_a(q) = (\ln \Sigma q_i^a)/(1 - a)$ - which was introduced by Rényi (1961) in the theory of information. Indeed, it can be proved that $H_a(q)= \ln(N_a(q))$ (cf. Hill, *ibid.*). It has been suggested that $H_a(q)$ can be used as a diversity measure (Pielou, 1975, p. 8).

[9] Although this assumption appears to be universally valid, Bhargava and Doyle (1974, p. 243) suggest that in certain cases the "even" population vector may be different from **1/k**.

[10] This follows from the equalities $D_2(\mathbf{1/k},q) = \Sigma(q_i - 1/k)^2 = \Sigma q_i^2 - 1/k = C(q) - 1/k$.

[11] Cf. Rao (1982, p. 275). Note that *Div(q) cannot* be directly defined as the distance between *q* and *the* maximally homogeneous vector since there are *k* maximally homogeneous vectors: (1,0,...,0), (0,1,0,...,0), ... ,(0,...,0,1).

[12] Note that "this interpretation strictly applies if the population size is infinite or if the sampling is done with replacement" (Agresti and Agresti, 1977, p. 206). More generally, this interpretation of $G(q)$ applies to any multivariate Bernoulli process. Probabilistic explications of Gini diversity have been suggested by Greenberg (1956) and Lieberson (1969, p. 851) and others.

[13] To describe such encounters some theoretical model including appropriate idealizations should be used. For instance, when dealing with animal communities, 'disturbing factors' such as "spatial patchiness and clumping in species distribution, differential mobility" and so on should be ignored (Sugihara, 1982, p. 565).

[14] Patil and Taillie (1979, pp. 7-11; 1982 pp. 549-550) develop a somewhat general encounter-theoretic explication of diversity and show that most 'traditional' diversity indices have plausible encounter-theoretic interpretations.

[15] For example, when the population is an animal community then its natural habitat might be one such entity.

[16] Similar remarks apply to any other quantity - such as the mean, variance, covariance - considered in statistics and, more generally, also apply to any scientific concept.

[17] Sugihara (1982, p. 565), referring to the diversity measures used in the biology of populations, remarks that "truly ground-breaking contributions to the theory of species diversity are not likely to arise in vitro from a mathematical analysis of indices but will most probably depend on an interplay of analysis with real data."

[18] This occurs, for instance, in a human community whose members are classified relative to a qualitative variable such as religion, ethnic origin, political party, etc.

[19] Further references on the sociological and economical application of diversity measures are given by Sen (1973) and Agresti and Agresti (1977).

[20] Patil and Taillie (*ibid.*) provide several references on these arguments.

[21] For instance, Patil and Taillie (1982, p. 565) report that the Committee to Review Methods for Ecotoxicology of the Environmental Sciences Board of the U.S. National Academy of Sciences (1981) concluded that "diversity is a system property that is likely to be a sensitive measure of ecosystem contamination."

[22] See, for instance, Taillie (1979, Section IV).

CHAPTER 11

[1] This reflects a limitation common to most studies on scientific enterprise. With reference to this, Laudan (1983, p. 2) points out that "students of the development of science, whether sociologists or philosophers, have alternatively been preoccupied with explaining consensus in science or with highlighting disagreement and divergence. Those contrasting focuses would be harmless if all they represented were differences of emphasis or interest. ... What creates tension is that neither approach has shown itself to have the explanatory resources for dealing with both."

[2] Niiniluoto (1987, p. 473) points out that "measures of verisimilitude always have a pragmatic dimension, by being dependent on our cognitive interests". More generally, he remarks that "science is a fallible and progressive enterprise which is run by historically developing scientific communities. Our tools for analyzing science should be flexible enough to take into account this richness of scientific practice." Other interesting remarks on the context-dependence of verisimilitude measures are made by Niiniluoto (*ibid.*, Chapter 13.4).

[3] For instance, one might look for a contextual justification of certain mixtures of Dirichlet distributions which have been suggested as appropriate priors for some kinds of multinomial contexts: see Good (1965, 1983), Dalal and Hall (1983), and Skyrms (1993?). In particular, Skyrms (1993?) shows that certain Dirichlet mixtures can be used to build exchangeable inductive methods capable of dealing with the so-called analogy by similarity (cf. Chapter 6.1). Moreover, one might look for a contextual justification of certain exchangeable inductive methods which attribute a positive probability to generalizations: see Hintikka (1966), Hintikka and Niiniluoto (1976), Kuipers (1978, Chapter 6) and Jamison (1970, pp. 50-53). In particular, Jamison (*ibid.*) specifies the prior distribution (on a parameter vector $\mathbf{q}$) equivalent to Hintikka's inductive methods.

[4] Several Bayesian statistical analyses of Markov processes have been proposed, e.g., Martin (1967). The problems concerning the inductive inferences relative to Markov processes and other 'non-Bernoulli' multicategorical processes have also been considered within the conceptual framework of TIP: see Achinstein (1963), Carnap (1963), Diaconis and Freedman (1980), Kuipers (1988), Skyrms (1991).

[5] More generally, quantitative processes include all those experimental processes where the result of a trial is the value of a given quantity.

[6] For instance, a CC-solution to EPO might be based on an external estimate of the entropy of the multivariate Bernoulli process under consideration (cf. Chapter 9, note 13).

[7] Note that various measures of quantitative variability have been defined in statistics and in some empirical sciences (see, for instance, formula (10.1)).

[8] Moreover, an interesting problem concerns the analysis of LPO in those cases where the inductive conclusion of a Bayesian inference is equated with the posterior distribution on the parameter (vector). Of course a necessary presupposition for such analysis is the definition of an appropriate measure of the distance between any possible posterior distribution and any possible value of the parameter (vector).

[9] In Chapter 8.6 the equivalence between the CC-solution and the V-solution to EPO was proved and explained. The existence of similar relationships between alternative V-solutions to EPO and alternative CC-solutions remains an open question.

REFERENCES

Achinstein, P., 'Confirmation Theory, Order and Periodicity', *Philosophy of Science* **30** (1963), pp. 17-35.

Agresti, A. and Agresti, B. F., 'Statistical Analysis of Qualitative Variation', in K. F. Schuessler (ed.), *Sociological Methodology 1978*, 1977, pp. 204-237.

Aitchison, J. and Dunsmore, I. R., *Statistical Prediction Analysis*, Cambridge University Press, Cambridge, 1975.

Amemiya, E. C., 'Measurement of Economic Differentiation', *Journal of Regional Science* **5** (1963), pp. 85-87.

Bar-Hillel, Y., 'The Acceptance Syndrome', in Lakatos (1968), pp. 150-161.

Barnett, V., *Comparative Statistical Inference*, Wiley, New York, 1973.

Bernardo, J. M., 'Reference Posterior Distributions for Bayesian Inference' (with Discussion), *JRSS Ser. B* **41** (1979), pp. 113-147.

Bhargava, T. N. and Doyle, P. H., 'A Geometric Study of Diversity', *Journal of Theoretical Biology* **43** (1974), pp. 241-251.

Bhargava, T. N. and Uppuluri, V. R. R., 'On an Axiomatic Derivation of Gini Diversity, with Applications', *Metron* **33** (1975), pp. 41-53.

Bogdan, R. J. (ed.), *Local Induction*. D. Reidel, Dordrecht, 1976.

Box, G. and Tiao, G., *Bayesian Inference in Statistical Analysis*, Addison-Wesley, Reading, Mass., 1973.

Brink, C., 'Verisimilitude: Views and Reviews', *History and Philosophy of Logic*, **10** (1989), pp. 181-201.

Burks, A. W., 'On the Significance of Carnap's System of Inductive Logic for the Philosophy of Induction, in P. A. Schilpp (1963), pp. 739-759.

Carnap, R., *The Logical Foundations of Probability*, The University of Chicago Press, Chicago, 1950. (2nd enlarged ed. 1962.)

Carnap, R., *The Continuum of Inductive Methods*, The University of Chicago Press, Chicago, 1952.

Carnap, R., 'Variety, Analogy and Periodicity in Inductive Logic', *Philosophy of Science* **30** (1963), pp. 222-227.

Carnap, R., 'On Rules of Acceptance', in Lakatos (1968), pp. 146-150, (1968a).

Carnap, R., 'Inductive Logic and Inductive Intuition', in Lakatos (1968), pp. 258-267, (1968b).

Carnap, R., 'A Basic System of Inductive Logic, Part I', in Carnap and Jeffrey (1971), pp. 33-165.

Carnap, R., 'A Basic System of Inductive Logic, Part II', in Jeffrey (1980), pp. 7-155.

Carnap, R. and Jeffrey, R. C. (eds.), *Studies in Inductive Logic and Probability*, vol. I, University of California Press, Berkeley, 1971.

Carnap, R. and Stegmüller, W., *Induktive Logic und Wahrscheinlichkeit*, Springer-Verlag, Wien, 1959.

Connor, R. J. and Mosimann, J. E., 'Concepts of Independence for Proportions with a Generalization of the Dirichlet Distribution', *JASA* **64** (1969), pp. 194-206.

Costantini, D., 'Considerazioni sulle Ipotesi Profonde della Distribuzione Beta', *Statistica* **38** (1978), pp. 313-322.

Costantini, D., 'The Relevance Quotient', *Erkenntnis* **14** (1979), pp. 149-157.

Costantini, D., 'The Role of Inductive Logic in Statistical Inference' (with Discussion), *Epistemologia*, **7** (1984), pp. 153-170.

Dalal, S. R. and Hall, W. J., 'Approximating Priors by Mixtures of Natural Conjugate Priors', *JRSS Ser. B* **45** (1983), pp. 278-286.

Darroch J. N. and Ratcliff, D., 'A Characterization of the Dirichlet distribution', *JASA* **66** (1971), pp. 641-643.

David, H. A., 'Gini's Mean Difference Rediscovered', *Biometrika* **55** (1968), pp. 573-575.

David, H. A., 'Gini's Mean Difference', in S. Kotz and N. L. Johnson (eds.), *Encyclopaedia of Statistical Sciences, Vol. 3*, Wiley, New York, 1983, pp. 436-437.

de Finetti, B., 'La prévision: ses lois logiques, ses sources subjectives', *Ann. Inst. Henri Poincaré* **7** (1937), pp. 1-68. [English translation in Kyburg H. E. Jr. and Smokler, H. (eds.), *Studies in Subjective Probability*, Wiley, New York, 1964, pp. 93-158.]

Dennis, B., Patil, G. P., Rossi, O., Stehman S. and Taillie, C., 'A Bibliography of Literature on Ecological Diversity and Related Methodology', in Grassle, Patil, Smith, and Taillie (1979), pp. 319-353.

Descartes, R., *Rules for the Direction of the Mind* (ca. 1628), in *Philosophical Works*, vol. I (translated by E. S. Haldane and G. R. T. Ross, 1911).

Diaconis, P. and Freedman, D., 'De Finetti's Theorem for Markov Chains', *Annals of Probability* **8** (1980), pp. 115-130.

Doksum, K., 'Tailfree and Neutral Random Probabilities and Their Posterior Distributions', *The Annals of Probability* **2** (1974), pp. 183-201.

Domotor, Z., 'Probability Kinematics, Conditionals, and Entropy Principles', *Synthese* **63** (1985), pp. 75-114.

Earman, J., *Bayes or Bust? A Critical Examination of Bayesian Confirmation Theory*, The MIT Press, Cambridge, Mass., 1992

Fabius, J., 'Two Characterizations of the Dirichlet distributions', *Annals of Mathematical Statistics* **1** (1973), pp. 583-587.

Feller, W., *An Introduction to Probability Theory and Its Applications, Vol. I*, 2nd. ed., Wiley, New York, 1968[2].

Ferguson, T. S., *Mathematical Statistics: A Decision-Theoretic Approach*, Academic Press, New York, 1967.

Festa, R., manuscript, 1982.

Festa, R., 'Inferenza predittiva e misure di eterogeneita', in *Statistica* **44** (1984), pp. 111-129.

Festa, R., 'Epistemic Utilities, Verisimilitude, and Inductive Acceptance of Interval Hypotheses', in *Abstracts of the 7th International Congress of Logic, Methodology and Philosophy of Science*, Vol. I, Saltzburg, 1983, pp. 212-215.

Festa, R., 'A Measure for the Distance Between an Interval Hypothesis and the Truth', *Synthese* **67** (1986), pp. 273-320.

Festa, R., 'Theory of Similarity, Similarity of Theories, and Verisimilitude', in T. K. Kuipers (1987a), 1987, pp. 145-176, (1987a).

Festa, R., 'New Aspects of Carnap's Optimum Inductive Method', in *Abstracts of the 8th International Congress of Logic, Methodology and Philosophy of Science, Vol I*, Moscow, 1987, pp. 397-400, (1987b).

Festa, R. and Buttasi, C., 'Generalized Carnapian Systems, Dirichlet Distributions, and the Epistemological Problem of Optimality', in *Abstracts of the 8th International Congress of Logic, Methodology and Philosophy of Science, Vol I*, Moscow, 1987, pp. 390-393.

Fine, T., *Theories of Probability*, Academic Press, New York, 1973.

Friedman, M., 'Truth and Confirmation', *The Journal of Philosophy* **76** (1979), pp. 361-382.

Gaifman, H., 'Applications of de Finetti's Theorem to Inductive Logic', in Carnap and Jeffrey (1971), pp. 235-251.

Gaifman, H., 'Towards a Unified Concept of Probability', in B. Marcus *et al.* (eds.), *Logic, Methodology and Philosophy of Science*, North-Holland, Amsterdam, 1986, pp. 319-350.

Gatlin, L., *Information Theory and the Living System*, Columbia University Press, New York, 1972.

Giere. R. N., Review of Carnap and Jeffrey (1971), *Synthese* **31** (1975), pp. 187-199.

Gini, C., 'Considerazioni sulle probabilità a posteriori e applicazioni al rapporto dei sessi nelle nascite umane', *Studi Economico-Giuridici della Facoltà di Giurisprudenza dell'Università di Cagliari* **3** (1911). [Reprinted in *Metron* **15** (1949), pp. 1-37.]

Gini, C., 'Variabilità e Mutabilità', *Studi Economico-Giuridici della Facoltà di Giurisprudenza dell'Università di Cagliari* **3** (1912), pp. **3-159**. [Reprinted, with few variations, in C. Gini, *Memorie di metodologia statistica*, vol. I, Veschi, Roma, 1955, pp. 211-282.]

Good, I. G., *Probability and the Weighing of Evidence*, Charles Griffin and Co., London, 1950.

Good, I. J, 'Maximum Entropy for Hypotheses Formulation, especially for Multi-dimensional Contingency Tables', *Annals of Mathematical Statistics* **34** (1963), pp. 911-934.

Good, I. J., *The Estimation of Probabilities*, The MIT Press, Cambridge, Mass., 1965.

Good, I. J., 'Comment' on Patil and Taillie (1982), pp. 551-563.

Good, I. J., *Good Thinking*, The University of Minnesota Press, Minneapolis, 1983, (1983a).

Good, I. J., 'The Robustness of a Hierarchical Model for Multinomials and Contingency Tables', in G. E. P. Box *et al.* (eds.), *Scientific Inference, Data Analysis, and Robustness*, 1983, pp. 191-211, (1983b).

Grassle, J. F., Patil, G. P., Smith, W., and Taillie, C. (eds.), *Ecological Diversity in Theory and Practice*, International Co-operative Publishing House, Burtonsville, 1979.

Graves, J., 'Uniformity and Induction', *British Journal for Philosophy of Science* **24** (1974), pp. 301-318.

Greenberg, J. H., 'The Measurement of Linguistic Diversity', *Language* **32** (1956), pp. 109-115.

Guiasu, R., *Information Theory with Applications*, McGraw-Hill, New York, 1977.

Guiraud, P., *Problèmes et Méthodes de la Statistique Linguistique*, D. Reidel, Dordrecht, 1959.

Hacking, I., *The Emergence of Probability*, Cambridge University Press, Cambridge, 1975.

Hardy, G. F., correspondence in *Insurance Records* **27** (1889). [Reprinted in *Trans. Fac. Actuar.* **8** (1929), pp. 174-182.]

Harsanyi, J. C., 'Acceptance of Empirical Statements: a Bayesian Theory Without Cognitive Utilities', Theory and Decision **18** (1985) pp. 1-30.

Hartigan, J., 'Invariant Prior Distributions', *Annals of Mathematical Statistics* **35** (1964), pp. 836-845.

Hempel, C. G., 'Deductive-nomological versus Statistical Explanations', in H. Feigl and G. Maxwell (eds.), *Scientific Explanation, Space and Time*, Minnesota Studies in the Philosophy of Science **3**, University of Minnesota Press, Minneapolis, 1962, pp. 98-169.

Herdan, G., *Quantitative Linguistics*, Butterworth, Whashington, D.C, 1964.

Herdan, G., *The Advanced Theory of Language as Choice and Chance*, Springer-Verlag, New York, 1966.

Hill, M. O., 'Diversity and Evenness: a Unifying Notation and its Consequences', *Ecology* **54** (1973), pp. 427-432.

Hilpinen, R., *Rules of Acceptance and Inductive Logic*, North-Holland, Amsterdam, 1968.

Hilpinen, R., 'Carnap's New System of Inductive Logic', *Synthese* **25** (1973), pp. 307-333.

Hintikka, J., 'A Two-dimensional Continuum of Inductive Methods', in Hintikka and Suppes (1966), pp. 113-132.

Hintikka, J., 'Unknown probability, Bayesianism, and de Finetti's representation theorem', in R. C. Buck and R. S. Cohen (eds.), *In Memory of Rudolf Carnap*, D. Reidel, Dordrecht, pp. 325-341.

Hintikka, J., 'Replies and Comments', in R. J. Bogdan (ed.), *Jakko Hintikka*, D. Reidel, Dordrecht, 1987, pp. 277-344.

Hintikka, J. and Niiniluoto, I., 'An Axiomatic Foundation of Inductive Generalization', in Przełecki, M. *et al.* (eds.), *Formal Methods in the Methodology of Empirical Sciences*, 1976, pp. 57-81.

Hintikka, J. and Pietarinen, J., 'Semantic Information and Inductive Logic', in Hintikka and Suppes (1966), pp. 96-112.

Hintikka, J. and Suppes, P., (eds.), *Aspects of Inductive Logic*, D. Reidel, Dordrecht, 1966.

Hintikka, J. and Suppes P., (eds.), *Information and Inference*, D. Reidel, Dordrecht, 1970.

Hobbes, T., *Humane Nature*, 1650.

Hogarth, R. M., *Judgment and Choice: the Psychology of Decision*, Chichester, Wiley, 1980.

Horwich, P., *Probability and Evidence*, Cambridge University Press, Cambridge, 1982.

Howson, C. and Urbach, P., *Scientific Reasoning. The Bayesian Approach*, Open Court, La Salle, Illinois, 1989.

Hume, D., *A Treatise of Humane Nature*, 1739.

Hurlbert, S. H., 'The Nonconcept of Species Diversity: a Critique and Alternative Parameters', *Ecology* **52** (1971), pp. 577-586.

Huzurbazar, V. S. *Sufficient Statistics*, Marcel Dekker, New York, 1976.

Huzurbazar, V. S., 'Bayesian Inference and Invariant Prior Probabilities', in Zellner (1980), pp. 445-449.

James, I. R. and Mosimann, J. E., 'A New Characterization of the Dirichlet Distribution through Neutrality', *The Annals of Statistics* **8** (1980), pp. 183-189.

Jamison, D., 'Bayesian Information Usage', in Hintikka and Suppes (1970), pp. 28-57.

Jaynes, E. T., 'Prior Probabilities', *IEEE Transactions on Systems Science and Cybernetics* SSC-4, pp. 227-241.

Jaynes, E. T., 'Where Do we Stand with Maximum Entropy?', in R. D. Levine and M. Tribus (eds.), *The Maximum Entropy Formalism*, The MIT Press, Cambridge, Mass., 1979, pp. 15-118.

Jaynes, E. T., *Papers on Probability, Statistics, and Statistical Physics*, (ed. by R. Rosenkrantz), D. Reidel, Dordrecht, 1983.

Jeffrey, R., 'Valuation and Acceptance of Scientific Hypotheses', *Philosophy of Science* **23** (1956), pp. 237-246.

Jeffrey, R., 'Probable Knowledge', in Lakatos (1968), pp. 166-180.

Jeffrey, R., 'Dracula Meets Woolfman: Acceptance vs. Partial', in M. Swain (ed.), *Induction, Acceptance, and Rational Belief*, D. Reidel, Dordrecht, 1970, pp. 157-185.

Jeffrey, R., 'Probability Measures and Integrals', in Carnap and Jeffrey (1971), pp. 167-223.

Jeffrey, R., 'Carnap's Inductive Logic', *Synthese* **25** (1973), pp. 299-306.

Jeffrey, R. (ed.), *Studies in Inductive Logic and Probability*, vol. II, University of California Press, Berkeley, 1980.

Jeffrey, R., 'De Finetti's Probabilism', *Synthese* **60** (1984), pp. 73-90.

Jeffreys, H., *Theory of Probability*, 2nd ed., Clarendon Press, Oxford, 1948 (3nd ed. 1961).

Johnson, W. E., 'Probability: Deductive and Inductive Problems (Appendix)', *Mind* **49** (1932), pp. 409-423.

Kaplan, M., 'A Bayesian Theory of Rational Acceptance', *The Journal of Philosophy* **78** (1981), pp. 305-330.

Kemeny, J., 'Carnap's Theory of Probability and Induction', in Schilpp (1963), pp. 711-738.

Kuipers, T. A. F., *Studies in Inductive Probability and Rational Expectation*, D. Reidel, Dordrecht, 1978.

Kuipers, T. A. F., 'A Survey of Inductive Systems', in Jeffrey (1980), pp. 183-192.

Kuipers, T. A. F., 'Approaching Descriptive and Theoretical Truth', *Erkenntnis* **18** (1982), pp. 343-378.

Kuipers, T. A. F., 'An Approximation of Carnap's Optimum Estimation Method', *Synthese* **61** (1984), pp. 361-362.

Kuipers, T. A. F., 'Some Estimates of the Optimum Inductive Method', *Synthese* **24** (1986), pp. 37-46.

Kuipers, T. A. F. (ed.), *What is Closer-to-the-truth?*, Rodopi, Amsterdam, 1987, (1987a).

Kuipers, T. A. F., 'A Structuralist Approach to Truthlikeness', in Kuipers (1987a), pp. 79-99, (1987b).

Kuipers, T. A. F., 'Inductive Analogy by Similarity and Proximity', in D. H. Helman (ed.), *Analogical Reasoning*, Dordrecht, Kluwer, 1988.

Kuipers, T. A. F., 'Naive and Refined Truth Approximation', *Synthese* **93** (1992), pp. 299-342.

Kullback, S., *Information Theory and Statistics*, Wiley, New York, 1959.

Kullback, S. and Leibler, R. A., 'On Information and Sufficiency', *Annals of Mathematical Statistics* **22** (1951), pp. 79-86.

Kyburg, H. E. Jr., *The Logical Foundations of Statistical Inference*, D. Reidel, Dordrecht, 1974.

Lakatos, I., *The Problem of Inductive Logic*, North-Holland Publishing Company, Amsterdam, 1968.

Laplace, P. S., 'Mémoire sur la probabilité des causes par les évenements', *Mémoires de l'Academie Royale des sciences presentés par divers savans*, **6** (1774), pp. 621-656. [Reprinted in *Oeuvres complète de Laplace*, vol. 7, Gauthiers-Paris. Translated in Stigler, S. M., 'Laplace's Memoir on Inverse Probability', *Statistical Science*, **1** (1986), pp. 359-378.]

Laudan, L., 'Peirce and the Trivialization of Self-Correcting Thesis', in R. N. Giere and R. S. Westfall (eds.), *Foundations of Scientific Method: the Nineteenth Century*, 1973, Indiana University Press, pp. 275-306.

Laudan, L., *Science and Values*, University of California Press, 1984.

La Valle, I.H., *An Introduction to Probability, Decision and Inference*, Holt, Rinehart and Winston, New York, 1970.

Levi, I., *Gambling with Truth*, Alfred A. Knopf, New York, 1967.

Levi, I., 'Acceptance Revisited', in R. Bogdan (ed.), *Local Induction*, D. Reidel, Dordrecht, 1976, pp. 1-71.

Levi, I., *The Enterprise of Knowledge: An Essay on Knowledge, Credal Probability, and Chance*, The MIT Press, Cambridge, Mass. and London, 1980.

Lewis, D., 'Immodest Inductive Methods', *Philosophy of Science* **38** (1971), pp. 54-63.

Lewis, D., 'Spielman and Lewis on Inductive Immodesty', *Philosophy of Science* **41** (1974), pp. 84-85.

Lewontin, R. C., 'The Apportionment of Human Diversity', *Evolutionary Biology* **6** (1972), pp. 381-398.

Lidstone, G. J., 'Note on the General Case of the Bayes-Laplace Formula for Inductive or Posterior probabilities', *Trans. Fac. Actuar.* **8** (1920), pp. 182-192.

Lieberson, S., 'Measuring Population Diversity', *American Sociological Review* **34** (1969), pp. 850-862.

Lindgren, B. W., *Statistical Theory*, MacMillan Publishing Co., New York, 1976[2].

Lindley, D. V., 'On a Measure of Information Provided by an Experiment', *Annals of Mathematical Statistics* **29** (1956), pp. 986-1005.

Lindley, D. V., *Bayesian Statistics, A Review*, S.I.A.M., Philadelphia, 1972.

Lourens, P. F., *The Formalization of Knowledge by Specification of Subjective Probability Distributions*, Dissertation, University of Groningen, 1984.

Martin, J. J., *Bayesian Decision Problems and Markov Chains*, Wiley, New York, 1967.

May, S., and Harper, W., 'Towards an Optimization Procedure for Applying Minimum Change Principles in Probability Kinematics', in W. L. Harper and C. A. Hooker (eds.), *Foundations of Probability Theory, Statistical Inference, and Statistical Theories of Science*, vol. I, D. Reidel, Dordrecht, 1976, pp. 137-166.

Mill, J. S., *A System of Logic*, London, 1843.

Miller, D., 'On the Distance from the Truth as a True Distance', in J. Hintikka *et al.* (eds.), *Essays on Mathematical and Philosophical Logic*, North-Holland, Amsterdam, 1978, pp. 415-435.

Miller, D., Review of Rosenkrantz (1977), *Philosophical Quarterly* **30** (1980), pp. 264-266.

Miller, D., 'Some Logical Mensuration', *The British Journal for the Philosophy of Science* **41** (1990), pp. 281-290.

Milne, P., 'A Note on Scale Invariance', *The British Journal for the Philosophy of Science* **34** (1983), pp. 49-55.

Mosimann, J. E., 'On the Compound Multinomial Distribution, the Multivariate ß-distribution, and Correlations among Proportions', *Biometrika* **49** (1962), pp. 65-82.

Niiniluoto, I., 'Scientific Progress', *Synthese* **45** (1980), pp. 427-462.

Niiniluoto, I., 'Inductive Logic as a Methodological Research Programme', *Scientia: Logic in the 20th Century*, Milano, 1983, pp. 77-100.

Niiniluoto, I., 'Truthlikeness and Bayesian Estimation', *Synthese* **67** (1986), pp. 321-346.

Niiniluoto, I., *Truthlikeness*, D. Reidel, Dordrecht, 1987.

Niiniluoto, I., 'Corroboration, Verisimilitude, and the Success of Science', in K. Gavroglu, Y. Goudaroulis and P. Nicolacopoulos (eds), *Imre Lakatos and Theories of Scientific Change*, 1989, pp. 229-243.

Novick, M. R. and Hall, W. J., 'A Bayesian Indifference Procedure', *JASA* **60** (1965), pp. 1104-117.

Novick, M. R. and Jackson, P. H., *Statistical Methods for Educational and Psychological Research*, McGraw Hill, New York, 1974.

Oddie, G., *Likeness to the Truth*, D. Reidel, Dordrecht, 1986.

Patil, G. P. and Taillie, C., 'An Overview of Diversity', in Grassle, Patil, Smith, and Taillie, (1979), pp. 3-27.

Patil, G. P. and Taillie, C., 'Diversity as a Concept and its Measurement', *JASA* **77** (1982), pp. 548-561.

Pearson, K., 'Mathematical Contributions to the Theory of Evolution. On a form of spurious correlation which may arise when indices are used in the measurement of organs', *Proceedings of Royal Society,* 1897, pp. 489-498.

Perks, W., 'Some Observations on Inverse Probability Including a New Indifference Rule' (with Discussion), *Journal of the Institute of Actuaries* **73** (1947), pp. 285-334.

Pielou, E. C., *An Introduction to Mathematical Ecology*, Wiley, New York, 1969.

Pielou, E. C., *Ecological Diversity*, Wiley, New York, 1975.

Pielou, E. C., *Mathematical Ecology*, Wiley, New York, 1977.

Pielou, E. C., 'Diversity Indices', in S. Kotz and N. L. Johnson (eds.), *Encyclopaedia of Statistical Sciences, Vol. 2*, Wiley, New York, 1982, pp. 408-412.

Pietarinen, J., *Lawlikeness, Analogy and Inductive Logic*, (Acta Philosophica Fennica **26**), North-Holland Publishing Co., Amsterdam, 1972.

Pietarinen, J., 'Inductive Immodesty and Lawlikeness', *Philosophy of Science* **41** (1974), pp. 196-198.

Popper, K. R., *Conjectures and Refutations: The Growth of Scientific Knowledge*, Routledge and Kegan Paul, London, 1963.

Rao, C. R., 'Gini-Simpson Index of Diversity: A Characterization, Generalization and Applications', *Utilitas Matematica* **218** (1982), pp. 273-282.

Rényi, A., 'On Measures of Entropy and information', *Proceedings of the Fourth Berkeley Symposium on Mathematical Statistics and Probability*, Vol. I, 1961, pp. 547-561.

Roberts, H. V., 'Probabilistic Prediction', *JASA* **60** (1965), pp. 50-62.

Rosenkrantz, R. D., *Inference, Method and Decision*, D. Reidel, Dordrecht, 1977.

Rosenkrantz, R. D., 'Measuring Truthlikeness', *Synthese* **45** (1980), pp. 463-487.

Rosenkrantz, R. D., *Foundations and Applications of Inductive Probability*, Ridgeview Publishing Company, Atascadero, California, 1981.

Rosenkrantz, R. D., 'Does the Philosophy of Induction Rest on a Mistake?', *Journal of Philosophy* **79** (1982), pp. 78-97.

Savage, L.J., *The Foundations of Statistics*, Wiley, New York, 1954 (2nd ed., Dover, New York, 1972.)

Schilpp, P. A., *The Philosophy of Rudolf Carnap*, Open Court, Lassalle, Illinois, 1963.

Schurz, G. and Weingartner, P., *Verisimilitude defined by relevant consequence-elements*, in Kuipers (1987a), pp. 47-77.

Seidenfeld, T., 'Why I am not an Objective Bayesian: some Reflections Prompted by Rosenkrantz', *Theory and Decision* **11** (1979), pp. 413-430, (1979a).

Seidenfeld, T., *Philosophical Problems of Statistical Inference. Learning from R. A. Fisher*, D. Reidel, Dordrecht, 1979, (1979b).

Seidenfeld, T., 'Entropy and Uncertainty', *Philosophy of Science* **53** (1986), pp. 467-491.

Sen, A., *On Economic Inequality*, Clarendon Press, Oxford, 1973.

Shannon, C. E., 'A Mathematical Theory of Communication', *Bell System Technical Journal* **27** (1948), pp. 379-423, 623-656. [Reprinted in C. E. Shannon and W. Weaver, *The Mathematical Theory of Communication*, University of Illinois Press, Urbana, Ill., 1949.]

Shimony, A., 'The Status of the Principle of Maximum Entropy', *Synthese* **63** (1985), pp. 35-53.

Simpson, E. H., 'Measurement of Diversity', *Nature* **163** (1949), pp. 688.

Skilling, J., 'Prior Probabilities', *Synthese* **63** (1985), pp. 1-34.

Skyrms, B., *Choice and Chance*, Wadsworth Publishing Company, Inc. Belmont, Cal., 1966.

Skyrms, B, 'Carnapian Inductive Logic for Markov Chains', *Erkenntnis* **35** (1991), pp. 439-460.

Skyrms, B., 'Analogy by Similarity in Hypercarnapian Inductive Logic', forthcoming, 1993?.

Spielman, S., 'Lewis on Immodest Inductive Methods', *Philosophy of Science* **39** (1972), pp. 375-377.

Spielman, S., Review of Rosenkrantz (1977), *The Journal of Philosophy* **78** (1981), pp. 356-367.

Sugihara, G., 'Comment' on Patil and Taillie (1982), pp. 564-565.

Swinburne, R., *An Introduction to Confirmation Theory*, Metheun & Co. Ltd, London, 1973.

Terlouw, P., *Subjective Probability Distributions: a Psychometric Approach*, Dissertation, University of Groningen, 1989.

Tintner, G.T., 'Foundations of Probability and Statistical Inference', (with discussion), *JRSS Ser. A* **112** (1949), pp. 251-286.

Villegas, C., 'On the Representation of Ignorance', *JASA* **72** (1977), pp. 650-654.

Walk, K., 'Kumulative Information', *Nachrichtentecnische Zeitschrift* **16** (1963), pp. 523-528.

Walk, K., 'Simplicity, Entropy and Inductive Logic', in Hintikka and Suppes (1966), pp. 66-80.

Wallace, A. R., *Tropical Nature and Other Essays*, MacMillan, London, 1875.

Watkins, J., 'The Popperian Approach to Scientific Knowledge', in G. Radnitzky and G. Andersson (eds.), *Progress and Rationality in Science*, D. Reidel, Dordrecht, 1978, pp. 23-44.

Weaver, W., 'Probability, Rarity, Interest, and Surprise', *The Scientific Monthly* **67** (1948), pp. 390-392.

Williams, P. M., 'Bayesian Conditionalization and the Principle of Minimum Information', *British Journal for the Philosophy of Science* **31** (1980), pp. 131-144.

Wilks, S. S., *Mathematical Statistics*, Wiley, New York, 1962.

Winkler, R. L., *Introduction to Bayesian Inference and Decision*, Holt, Rinehart and Winston, New York, 1972.

Winkler, R. L., 'Prior Information, Predictive Distributions, and Bayesian Model Building', in Zellner (1980), pp. 95-109.

Withworth, W. A., *DCC Exercises in Choice and Chance*, 1897 [Reprinted 1965, Hafner, New York].

Yule, G. U. *Statistical Study of Literary Vocabulary*, Cambridge University Press, London, 1944.

Zabell, S. L., 'W.E. Johnson's "Sufficientness" Postulate', *Annals of Statistics* **10** (1982), pp. 1091-1099.

Zabell, S. L., 'The Rule of Succession', *Erkenntnis* **31** (1989), pp. 283-321.

Zellner, A., *An Introduction to Bayesian Inference in Econometrics*, Wiley, New York, 1971.

Zellner, A. (ed.), *Bayesian Analysis in Econometrics and Statistics*, North-Holland Publishing Company, Amsterdam, 1980.

INDEX OF NAMES

INDEX OF SUBJECTS

LIST OF REQUIREMENTS AND ACRONYMS

Requirements of adequacy/principles/conditions/properties

Acronyms

223. A. García de la Sienra, *The Logical Foundations of the Marxian Theory of Value.* 1992 ISBN 0-7923-1778-5
224. D.S. Shwayder, *Statement and Referent.* An Inquiry into the Foundations of our Conceptual Order. 1992 ISBN 0-7923-1803-X
225. M. Rosen, *Problems of the Hegelian Dialectic.* Dialectic Reconstructed as a Logic of Human Reality. 1993 ISBN 0-7923-2047-6
226. P. Suppes, *Models and Methods in the Philosophy of Science: Selected Essays.* 1993 ISBN 0-7923-2211-8
227. R. M. Dancy (ed.), *Kant and Critique: New Essays in Honor of W. H. Werkmeister.* 1993 ISBN 0-7923-2244-4
228. J. Woleński (ed.), *Philosophical Logic in Poland.* 1993 ISBN 0-7923-2293-2
229. M. De Rijke (ed.), *Diamonds and Defaults.* Studies in Pure and Applied Intensional Logic. 1993 ISBN 0-7923-2342-4
230. B.K. Matilal and A. Chakrabarti (eds.), *Knowing from Words.* 1993 ISBN 0-7923-2345-9
231. S.A. Kleiner, *The Logic of Discovery.* A Theory of the Rationality of Scientific Research. 1993 ISBN 0-7923-2371-8
232. R. Festa, *Optimum Inductive Methods.* A Study in Inductive Probability, Bayesian Statistics, and Verisimilitude. 1993 ISBN 0-7923-2460-9

Previous volumes are still available.

KLUWER ACADEMIC PUBLISHERS – DORDRECHT / BOSTON / LONDON

MIX
Papier aus verantwortungsvollen Quellen
Paper from responsible sources
FSC® C105338

If you have any concerns about our products,
you can contact us on
ProductSafety@springernature.com

In case Publisher is established outside the EU,
the EU authorized representative is:
Springer Nature Customer Service Center GmbH
Europaplatz 3, 69115 Heidelberg, Germany

Printed by Libri Plureos GmbH
in Hamburg, Germany